D0481939

THE LITTLE BOOK OF
ORCHIDS
GEMS OF NATURE

Mark Chase is a senior research scientist at the Royal Botanic Gardens, Kew. He co-edited *Genera Orchidacearum* and has contributed to more than 500 publications on plant science.

Maarten Christenhusz is a botanist who has worked for the Finnish Museum of Natural History and the Royal Botanic Gardens, Kew. He is deputy editor of the *Botanical Journal of the Linnean Society*.

Tom Mirenda is the Director of Horticulture, Education, and Outreach at the Hawaii Tropical Botanic Garden and the former Orchid Collection Specialist at the Smithsonian Institution, Washington, D.C. He frequently lectures on orchid ecology and conservation in the US and abroad, and is a columnist for *Orchids*, the magazine of the American Orchid Society.

First published in the UK in 2020 by
Ivy Press
An imprint of The Quarto Group
The Old Brewery, 6 Blundell Street
London N7 9BH, United Kingdom
T (0)20 7700 6700
www.QuartoKnows.com

Material in this book was first published in 2017 in *The Book of Orchids*

British Library Cataloguing-in-Publication Data
A catalogue record for this book is available from the British Library

ISBN: 978-0-7112-5393-3

The views expressed in this work are those of the authors and do not necessarily reflect those of the publisher or the Royal Botanic Gardens, Kew.

This book was designed and produced by
Ivy Press
58 West Street, Brighton BN1 2RA, UK

Publisher David Breuer
Editorial Director Tom Kitch
Art Director James Lawrence
Commissioning Editors Stephanie Evans, Natalia Price-Cabrera
Project Editor Joanna Bentley
Designer Ginny Zeal
Map Artwork Richard Peters

Printed in China

10 9 8 7 6 5 4 3 2 1

THE LITTLE BOOK OF ORCHIDS

GEMS OF NATURE

MARK CHASE, MAARTEN CHRISTENHUSZ
& TOM MIRENDA

IVY PRESS

CONTENTS

INTRODUCTION

Unquestionably lovely, orchids are far beyond being just beautiful. They are seemingly endless in their diversity, perpetually compelling, and astonishingly well adapted to a mind-boggling array of ecological niches and evolutionary partners. A geologically old family, members of the Orchidaceae have colonized the far reaches of our planet save those most inhospitable: extreme poles, high mountain peaks, the most desolate deserts, and, of course, the deep waters of our lakes, rivers, and oceans.

Having evolved to occur in such a wide variety of habitats, as well as perfecting the ability to interact with and exploit myriad creatures as symbionts, orchids are the ideal plant family to teach us about biodiversity and illustrate its importance. The remarkable structures and colors of each and every orchid species convey a story about their ecology, evolution, and survival strategy. Once analyzed and unlocked, these stories give us powerful insight into the processes that have shaped our world for millennia and, hopefully, inspire us to conserve that which took millennia to create.

Masters of deception and manipulation, orchids are famous for lying and cheating their way to their many evolutionary successes. Exploring the manner in which they co-opt pre-existing behaviors of a bewildering cohort of pollinators of lilliputian dimensions is not only outstandingly instructive, but is just plain fun to contemplate. Even the venerable Charles Darwin referred to orchids as "Splendid Sport" and maintained a passion for them throughout his lifetime. It is undeniable that orchids have gripped the psyches of many humans. They have even, in recent years, become the most sold and cultivated type of ornamental plant. Their beauty alone does not explain this phenomenon.

Many theories exist as to why orchids are so alluring to us. It is thought that their zygomorphic (bilaterally symmetrical) flower structure influences us to see orchid flowers similarly to the way we see faces, attributing to them some "personality" in addition to their beauty. Some find the lip of certain orchids to be reminiscent of human anatomical parts that we normally keep covered, lending them a subliminal or feral attraction. Others simply find the combination of color, form, grace, and fragrance most appealing, yet not all orchids have traditionally attractive versions of these attributes. Some of the most compelling orchids are rank-smelling, muddy in coloration, and borne on clunky plants. Nothing adequately explains why people become so wildly obsessive about orchids. Ultimately, they are simply provocative creatures that manage to elicit strong reactions from pollinator and person alike.

In this book, we invite you to dip into the world of orchids and get to know these marvels of nature.

WHAT IS AN ORCHID?

The orchid family, Orchidaceae, embraces 26,000 species in 749 genera and is one of the two largest families of flowering plants, the other being that of the daisies and lettuce, Asteraceae. Many people have a vague idea of what an orchid is, but it is likely that most would not recognize all the species included in this book as orchids. So, what is an orchid?

Orchids are divided into five subfamilies, Apostasioideae, Vanilloideae, Cypripedioideae, Epidendroideae, and Orchidoideae. This subdivision is based on DNA studies and morphology and reflects major differences in vegetative features and especially in the way orchid flowers are constructed. The five subfamilies have been recognized in the past as separate families by some botanists based on these distinctive characteristics, and the only characteristic they all share is that of how orchid embryos develop, from a structure called a protocorm, which is a small ball of cells without roots, stems, or leaves.

To develop into a mature orchid plant, a protocorm has to be successfully infected by a fungus, from which the developing orchid seedling obtains initially all the food (in the form of sugars) and minerals it needs to grow. As they start their life, orchids can be thought of as parasites on fungi. However, most but not all orchids as adults go on to develop roots and leaves, and produce their own food through photosynthesis. At a much later stage the continuing relationship of an orchid plant with the fungus can become mutually beneficial. In nature, the orchid exchanges sugars produced by its photosynthesis for minerals found more effectively by the fungus. In cultivation, the need of an orchid protocorm for a fungal partner can be replaced by manufactured sources of food and minerals, and many orchids are grown commercially using germination media with added sugars and minerals.

The column

The other major trait that most botanists use to recognize an orchid is a structure called the gynostemium, or column, produced by the fusion of male (stamen) and female (stigma) parts in the flower. All but one of the five subfamilies share this feature. The exception is the subfamily Apostasioideae, consisting of only 14 species in two genera, Apostasia and Neuwiedia, which all lack complete fusion of the male and female parts.

The characteristically fused structure of the column, shared by 99.95 percent of all orchids, is responsible for the remarkable event where a pollinator, such as a bee, wasp, or moth, is maneuvered into doing exactly what the orchid wants. This allows the pollen, usually in the form of thousands of grains bound into a solid ball, or pollinium, to be placed on the animal in a precise manner and then, due to the close proximity of the stigma and anther (the part of the stamen holding the pollen), be precisely removed from that spot. Pollination in orchids is, therefore, a highly exact sequence of events, leading to fertilization of the thousands of developing orchid embryos in the carpel, or ovary, with just a single visit of a pollinator, provided that it has previously visited another flower of that same orchid species to pick up pollinia.

The lip

In most orchids the female receptive surface, or stigma, is a cavity on the side of the column that faces the other highly distinctive orchid structure: a modified petal (one of three) that is termed the labellum, or lip. This serves variously as a landing platform, a flag to attract the pollinator, or—playing an important part in various forms of deceit that orchids use to fool pollinators—a mimic of something the pollinator wants, such as nectar, pollen, or a mate.

There are many orchids that appear not to have a lip. A good example is the genus Thelymitra from Australia, where the member species are called sun orchids. Rather than a lip, the flowers of these plants have three sepals, which are initially a set of protective leaflike structures (that in many orchids also become colorful) and three similar petals (also colorful leaflike organs). Such similarity of all three petals, though, is the exception among orchids, most of which develop a highly modified lip.

Although it has long been known that orchids can control the appearance of the lip in isolation from the other showy parts of their flowers—the two remaining petals and three sepals—it was not clear until recently how the lip was controlled from a genetic or developmental perspective. In nearly all other plants that have been studied in this regard, the three petals are controlled by the same floral genes, and by and large they all three do the same thing and look the same.

Think, for example, of a lily or a tulip, in which the three petals are identical. In orchids, there has been a duplication of the floral genes, and one of the duplicated copies is expressed just in the lip, making it possible for this petal—the lip—to look different and be involved in pollinator manipulation apart from the other two petals, in which the gene is not expressed. This more complicated set of genetic controls has made the flowers of orchids among the most complex in the plant world and undoubtedly is a major reason why they are adapted for pollination by such a large range of animals.

Distinguishing features

The combination of column, lip, and pollinia—the first unique to orchids, the others not unique but unusual among plants—makes it possible for botanists to recognize plants as orchids despite their capacity to look decidedly un-orchidlike. In biological terms, this amalgam of features has enabled orchids to become evolutionarily explosive, leading to the 26,000 species alive today.

The Parts of an Orchid Flower

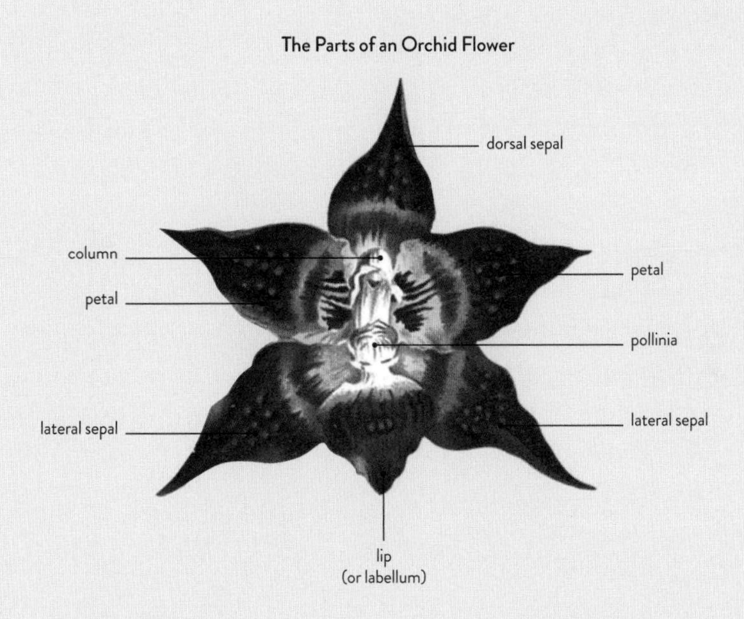

ORCHID EVOLUTION

Orchids evolved during the Late Cretaceous period, roughly 76 to 105 million years ago. This is much earlier than botanists once thought and makes Orchidaceae one of the 15 oldest angiosperm families, of which there are 416 in total. Few orchid fossils older than 20 to 30 million years have been found, and it was thought that orchids evolved relatively recently compared to many other groups of flowering plants. That they have a poor fossil record is not surprising because most orchids are herbs, which generally do not fossilize well, and their highly modified pollinia are difficult to recognize in the fossil record.

Dinosaur dependence

All five orchid subfamilies evolved before the end of the Cretaceous period, which means that orchids and dinosaurs overlapped. Considering the great diversity of orchid pollinators, we can only wonder if orchids managed to adapt to pollination by dinosaurs before the latter became extinct 65 million years ago. Vertebrates in general are uncommon orchid pollinators, and nearly all of those recorded are birds—direct descendants of the dinosaurs. There were many small species of dinosaurs, so it is possible that some visited flowers to collect nectar and, like many animals today, were deceived into pollinating orchids. Any orchids adapted to dinosaur pollination would have become extinct with their pollinator, and so are now lost to us.

Distribution

The discovery that orchids were much older than previously thought was a result of the widespread sequencing of DNA that only became possible in the mid-1990s. This greater age makes a good deal of sense when it comes to understanding the geographic distribution of orchids. It was long assumed that orchids could have reached their current worldwide distribution relatively recently by long-distance dispersal of their small, almost microscopic seeds. Due to their dependence for food and minerals on the fungi with which they associate, orchids do not include food reserves or minerals in their seeds, unlike, for example, a bean in which the stored food and minerals make up the bulk of its much larger seed. Orchid seeds are, therefore, light and easily distributed by the wind, which theoretically could propel them over long distances. However, the longer an orchid seed remains aloft, the more the small embryo dries out, making most orchid seeds inviable before they can travel great distances. So, most orchid species have a limited distribution, even as constrained as a single mountain. Orchids have instead achieved their worldwide distribution by passively riding the continents, which at the time the plants evolved were much closer than they are today.

POLLINATION

Orchids are well known for elaborate pollination mechanisms that have evolved to achieve the mating of different plants, or cross-fertilization. Flowers of most plants, including orchids, contain organs of both sexes, but self-pollination is as generally undesirable in plants as it is in animals. Most plants, and orchids in particular, have evolved methods, often exceedingly complicated, to avoid self-pollination happening. This process has long fascinated scientists, including Charles Darwin, who studied pollination of orchids in detail and was so enthralled by the plants that his first book after publication of *On the Origin of Species* (1859) was *Fertilization of Orchids*, entirely dedicated to orchids.

Pollinator deception

Most orchids produce pollen in two to six tight bundles, called pollinia. These are often attached to ancillary structures that together are called a pollinarium, which attaches the pollinia to the pollinator's body, usually in a position that makes it difficult for the animal to remove them. Most orchids look as if they contain a reward for pollinators but few actually offer it. Some even produce long nectar spurs that are devoid of nectar. Rates of visitation by pollinating insects to such deceptive flowers are, understandably, low. Insects learn quickly to avoid these rewardless flowers, but they make the mistake often enough for it to be effective in a system in which a single visit can result in deposition of thousands of pollen grains, each fertilizing one of the thousands of orchid ovules produced by each flower.

Darwin himself came to the conclusion that outcrossing, or pollination between unrelated plants, is so advantageous for most orchids that deceit and corresponding low rates of visitation are the general rule. Apparently, setting seeds in only a few flowers but guaranteeing that these are of high quality (due to cross-fertilization involving flowers on different plants) makes deceit a successful strategy. In this case, the cheating orchids have prospered, despite the fact that they treat the insects upon which they depend so badly. There is no mutual benefit for the orchid and its pollinators as there is in pollination systems with rewarding plants; the deceiving orchid could go extinct and the animal would only experience a slight improvement in its condition due to fewer floral visits without a reward. However, if the animal pollinating a deceitful orchid species becomes extinct, then the orchid also disappears or develops a method by which to self-pollinate its flowers, which has been known to evolve when an orchid species reaches an island without its pollinator accompanying it.

Seed production

The combination of delivery of whole pollinaria on a single visit and fertilization of a correspondingly large number of ovules in the ovary means that from a single pollinator visit a massive number of seeds can be produced. That many orchids, such as some species of Dendrobium, Epidendrum, and Oncidium, bear large inflorescences with hundreds of flowers may seem like an extreme waste of energy, but production of mature ovules ready for fertilization is delayed until pollination takes place, thus reducing energy inputs associated with these large numbers of flowers.

Mimicry and deceit

Deceit involving mimicry of other local plants that produce a reward for their pollinator is another common habit for orchids. Although not offering a reward itself, the orchid benefits from pollinators that fail to distinguish between a cheating orchid and the rewarding species, and so the former obtains a degree of pollinator service that drops dramatically if the latter is not present. In other cases, a deceitful orchid species is not mimicking a single reward-offering species in the immediate neighborhood, but rather is using a suite of the traits associated by pollinators with the presence of a reward. These include fragrance, color, "nectar guides" to direct a pollinator to the center of the flower, and a nectarless cavity or spur of the correct shape and size to suggest that nectar is present.

In many groups of orchids, a much more specific type of deceit, involving sexual attraction, has evolved. Darwin was unaware of this phenomenon, although he speculated on what might be happening with native British bee and fly orchids (genus Ophrys). The details would probably have shocked him and many other botanists of that time. It is thought that mimicry of the female of a species of bee, wasp, or fly begins as some other more general type of deceit and subsequently becomes more complicated and specific. For example, the orchid *Anacamptis papilionacea* appears not to be mimicking any specific nectar-producing species in its habitat and is instead just a general reward-flower mimic. However, there are more males than females among the insects it attracts, so it appears that some sort of sexual attraction is operating, which could lead to further change on the part of the orchid to enhance this aspect of the deceit.

Many orchids using visual sexual mimicry also produce floral fragrances that are identical to the sex pheromones produced by the female of the insect species to attract a male. This at first sounds wholly preposterous: how can a flower evolve to produce something so alien to a plant as an animal sex pheromone? However, once it became known how the biochemical pathways operate by which such animal hormones are produced, it also became clear that plants share these same general pathways and often produce minor amounts of such compounds as part of their general bouquet of scents. Thus, the assembly of a highly specific sexual pheromone starts out with production of small amounts of similar compounds that become predominant when an increased presence in the mixture generates higher rates of male visitation. When combined with visual cues, such fragrance compounds reinforce the "message" being sent to male insects, and sexual mimicry is the result. Orchids in many distantly related groups have independently evolved this sexual mimicry syndrome, which, now that we know the genetic and biochemical details, is not as surprising as it first appeared.

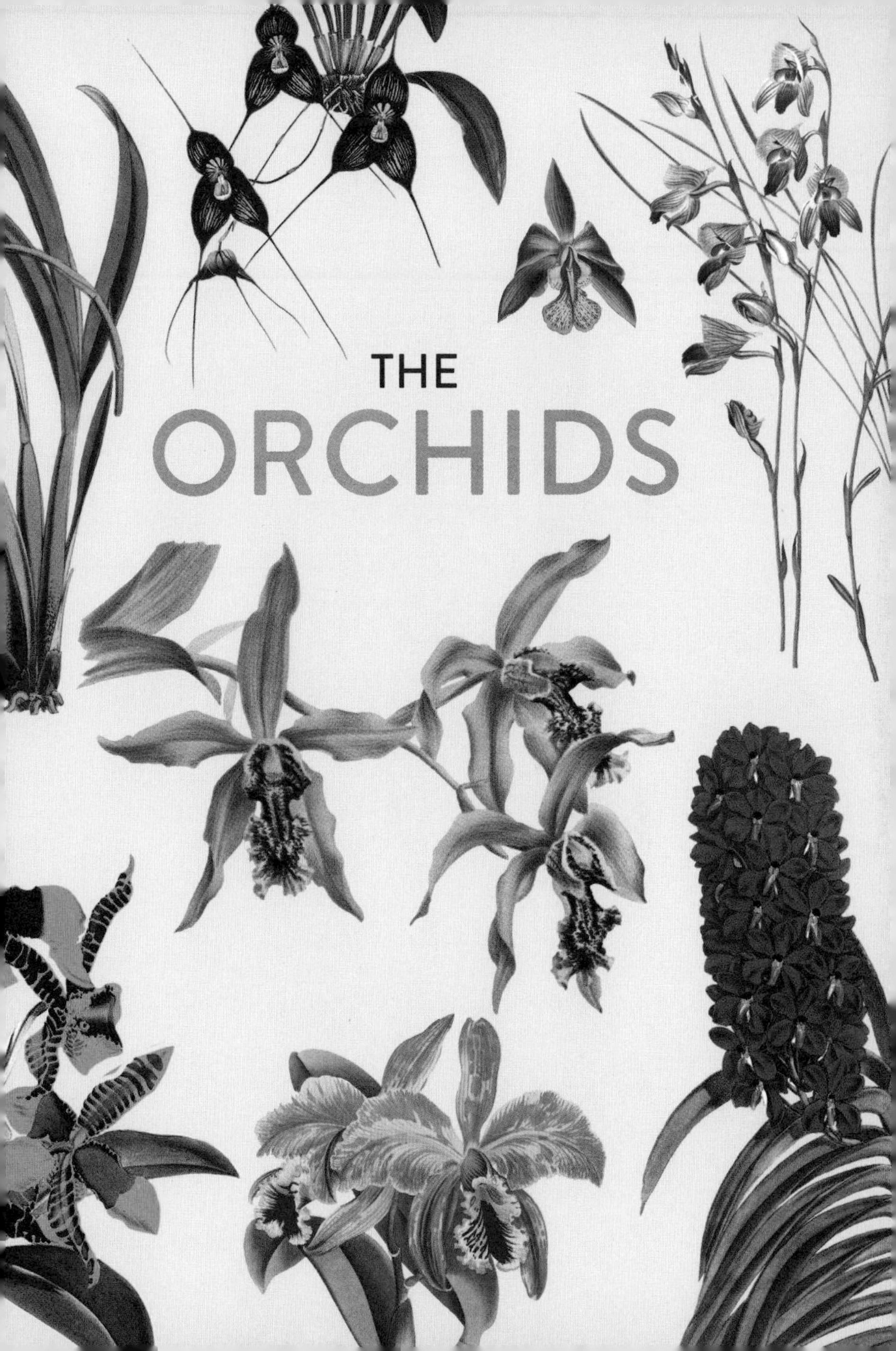

THE ORCHIDS

NEUWIEDIA VERATRIFOLIA

FALSE HELLEBORE ORCHID

BLUME, 1834

PLANT SIZE
22 × 18 in (56 × 46 cm)

FLOWER SIZE
1⅜ in (3.5 cm)

SUBFAMILY ~ Apostasioideae

TRIBE AND SUBTRIBE ~ Not applicable

NATIVE RANGE ~ Malesia to Melanesia, from Borneo and Java to the Philippines and Vanuatu, from sea level to 3,300 ft (1,000 m)

HABITAT ~ Evergreen dipterocarp forests on sandstone, limestone, ultramafic soil or shale, usually in deep shade under humid conditions

TYPE AND PLACEMENT ~ Terrestrial or on rocks

CONSERVATION STATUS ~ Locally abundant

FLOWERING TIME ~ June to September (wet season)

Few people, when they look at *Neuwiedia veratrifolia,* think it is an orchid. Named for German naturalist, ethnologist, and explorer Prince Maximilian Alexander Philipp zu Wied-Neuwied (1782–1867), these large, hairy plants produce up to ten plicate leaves that more closely resemble false hellebore (*Veratrum*, in the family Melanthiaceae). Like the genus *Apostasia*, *Neuwiedia* was placed in a separate family in the past because it has three free anthers instead of the single fused anther found in most other orchids. The two genera, however, share some unique traits with orchids and are now considered to be members of the Orchidaceae family.

Neuwiedia veratrifolia is self-compatible and mostly self-pollinating. In addition, stingless *Trigona* bees visit the flowers, vibrate the anthers, and are then dusted with the pollen released.

The flower of the False Hellebore Orchid has white crystals in its tissues. The upper sepals and petals are asymmetrical, while the lip is symmetrical and broader than the petals. Three stamens emerge from the column base, and the anthers are free from the style.

DISTRIBUTION

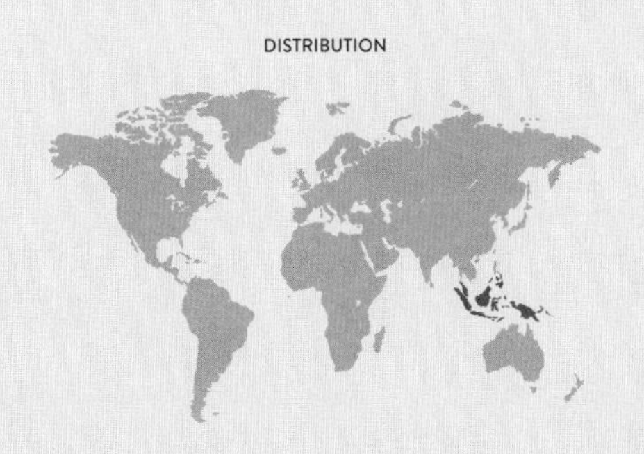

POGONIA OPHIOGLOSSOIDES

ROSE POGONIA

(LINNAEUS) KER GAWLER, 1816

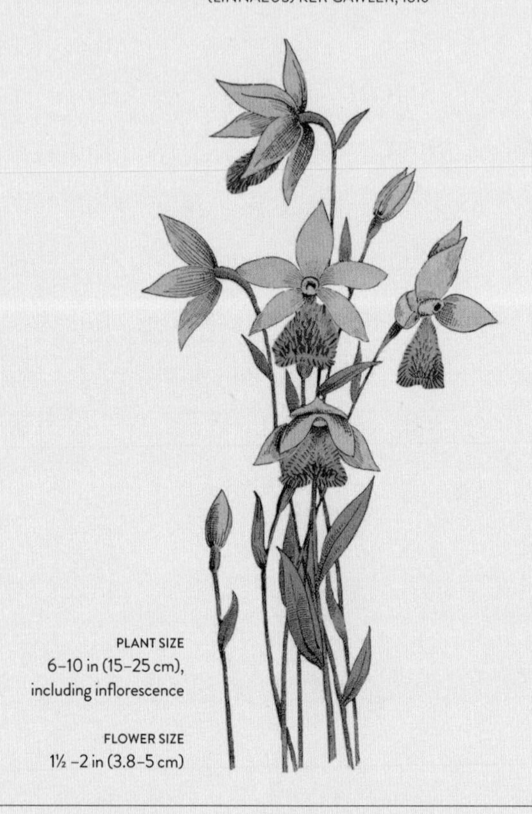

PLANT SIZE
6–10 in (15–25 cm), including inflorescence

FLOWER SIZE
1½ –2 in (3.8–5 cm)

SUBFAMILY ~ Vanilloideae

TRIBE ~ Pogonieae

NATIVE RANGE ~ Eastern North America, from Canada to Florida and west to Minnesota

HABITAT ~ Wet meadows, bogs, stream sides, often occurring in poorly drained roadside ditches

TYPE AND PLACEMENT ~ Terrestrial

CONSERVATION STATUS ~ Threatened or endangered

FLOWERING TIME ~ Early spring in south to midsummer in northern part of range

A slender, semi-aquatic plant, often occurring in bogs and beside streams, the pretty Rose Pogonia can be locally abundant, often proliferating into lush, multi-growth colonies. Preferring to grow where there is easily available, pure water, this species is scarce in years with sparse rainfall but will rebound in wet periods. The short-lived, mostly pale pink flowers can vary in color and intensity and probably use their darker fringed labellum with yellowish filamentous crests to attract pollinators. This open-jawed appearance explains the plant's alternative common names, Adder's Mouth or Snake Mouth. Underground, there is a mass of roots but no tuber.

Pogonias grow in dappled light, usually in moist sphagnum moss, and can produce massive colonies. The genus name comes from the Greek word *pogon*, meaning beard, which refers to the hairy labellum.

The flower of the Rose Pogonia is usually pale pink with a darker lip, fringed with purplish striations, and a yellow crest. Flowers appear singly on a stem, though up to three have been reported on vigorous plants.

DISTRIBUTION

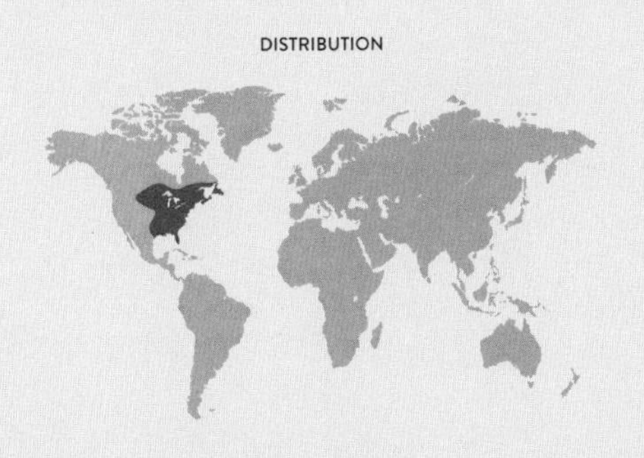

EPISTEPHIUM SCLEROPHYLLUM

LEATHER-LEAFED CROWN ORCHID

LINDLEY, 1840

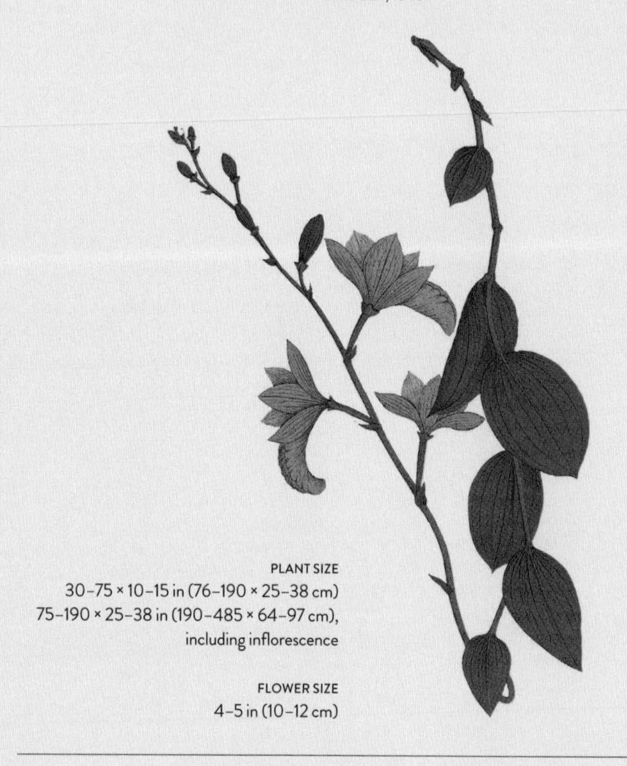

PLANT SIZE
30–75 × 10–15 in (76–190 × 25–38 cm)
75–190 × 25–38 in (190–485 × 64–97 cm), including inflorescence

FLOWER SIZE
4–5 in (10–12 cm)

SUBFAMILY ~ Vanilloideae

TRIBE ~ Vanilleae

NATIVE RANGE ~ Tropical South America, at 330–2,950 ft (100–900 m)

HABITAT ~ Open places in rain forests and savanna

TYPE AND PLACEMENT ~ Terrestrial

CONSERVATION STATUS ~ Not assessed

FLOWERING TIME ~ All year

This large, ground-dwelling orchid produces erect stems covered with leathery, rigid, ovate leaves, while underground there is a branching horizontal rhizome with many tough roots. The inflorescences are terminal and have small floral bracts with many flowers that open successively, two to three at a time. At the top of the ovary the flowers are inserted into a scalloped ridge. This crownlike structure is the basis of the genus name (Greek, *epi-*, "upon," and *stephanos*, "crown"). The plant is a member of the same tribe as the genus *Vanilla*, to which it is closely related.

The showy flowers have a classical orchid shape (like species of the genus *Cattleya*), which indicates that they are probably pollinated by bees. In spite of their fantastically beautiful flowers, these orchids have never been successfully cultivated.

The flower of the Leather-leafed Crown Orchid has three relatively narrow, pink sepals and two broader petals. The massive pink lip is wrapped around the column and has yellow and white nectar guide markings with a cluster of long hairs near its middle.

DISTRIBUTION

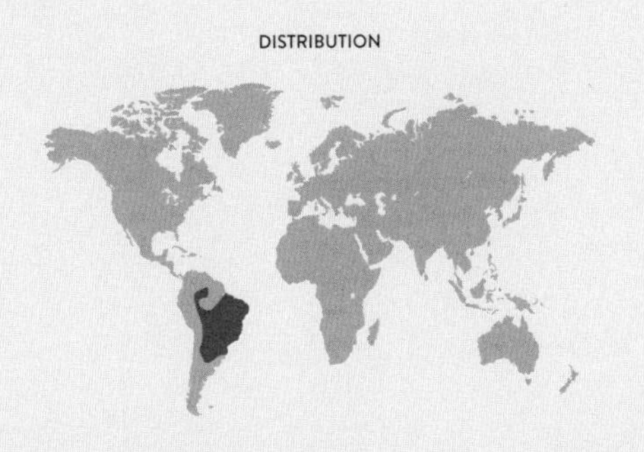

VANILLA PLANIFOLIA

VANILLA ORCHID

JACKSON EX ANDREWS, 1808

PLANT SIZE	FLOWER SIZE
20 ft (6 m) or more	2½ in (6.4 cm)

SUBFAMILY ~ Vanilloideae

TRIBE ~ Vanilleae

NATIVE RANGE ~ Mexico (but widely cultivated and naturalized elsewhere in the tropics)

HABITAT ~ Lowland tropical forests

TYPE AND PLACEMENT ~ Terrestrial, but climbs trees

CONSERVATION STATUS ~ Not threatened

FLOWERING TIME ~ Throughout the year

The most commercially important orchid, Vanilla is cultivated in tropical places around the world for its "beans," which, when dried and fermented, produce the popular flavoring. The genus *Vanilla* occurs on five continents, with more than a hundred species, and is one of only five vining orchid genera, which need the support of trees to grow to their full potential. The vines can grow to great lengths.

Short-lived flowers with tubular lips are produced successively on axial racemes. They are pollinated by a wasp in their natural range in Mexico, but pollination has to be done by hand in plantations in places such as Madagascar, Réunion, and Tahiti, where the plants are cultivated in large numbers. This one species provides 95 percent of the world's commercially produced vanilla pods.

The flower of the Vanilla Orchid is usually yellow or greenish, with similarly colored sepals, petals, and lip. Although flowers last a single day, the plants bloom frequently and successively over a long period.

DISTRIBUTION

CYPRIPEDIUM CALCEOLUS

YELLOW LADY'S SLIPPER

LINNAEUS, 1753

PLANT SIZE
15–30 in (38–76 cm), including flowers

FLOWER SIZE
2–3 in (5–8 cm)

SUBFAMILY ~ Cypripedioideae

TRIBE AND SUBTRIBE ~ Not applicable

NATIVE RANGE ~ Temperate northern Europe and Asia, from the British Isles to Korea and Japan

HABITAT ~ Temperate woodlands and scrub, at up to 6,600 ft (2,000 m)

TYPE AND PLACEMENT ~ Terrestrial

CONSERVATION STATUS ~ Widespread, although endangered in places

FLOWERING TIME ~ April to June (spring)

The Yellow Lady's Slipper has a vast native range over expansive areas of the Northern Hemisphere. Some have considered the North American species *Cypripedium parviflorum* to be the same or merely a varietal form of this beautiful orchid, but it is now known to be a distinct species. It thrives on damp substrate in limestone-rich areas, which may be why the species is amenable to cultivation. Despite the plant's widespread distribution, poaching and urban sprawl threaten some populations.

In the species name, *calceolus* (Latin) means "little shoe," and it is the slipperlike shape of the lip that has inspired both the scientific and common names. The pouch acts as a temporary insect trap, waylaying hapless pollinators, usually bees, but with no reward for their services. They are released with pollinia attached.

The flower of the Yellow Lady's Slipper varies but generally has yellow to brown sepals and petals, with a brilliant-yellow, pouch-shaped lip. Usually solitary flowers appear at the apex of a pubescent stem and are subtended by a leaflike bract.

DISTRIBUTION

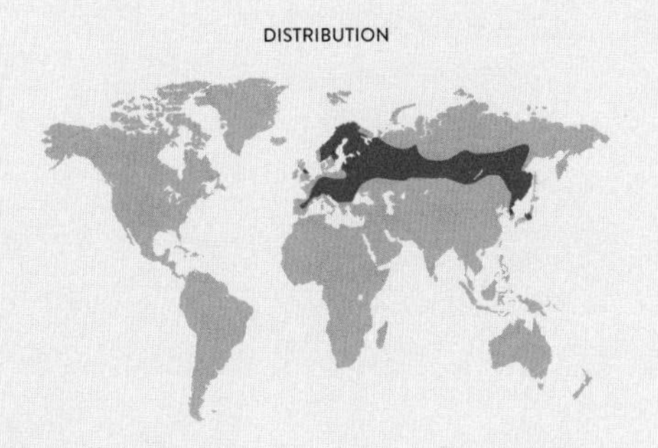

PAPHIOPEDILUM ROTHSCHILDIANUM

ROTHSCHILD'S SLIPPER ORCHID

(REICHENBACH FILS) STEIN, 1892

PLANT SIZE
10–15 × 12–20 in
(25–38 × 30–51 cm),
excluding inflorescence

FLOWER SIZE
6–10 in (15–25 cm)

SUBFAMILY ~ Cypripedioideae

TRIBE AND SUBTRIBE ~ Not applicable

NATIVE RANGE ~ Rain forests around Mount Kinabalu in northern Borneo, at 1,640–3,950 ft (500–1,200 m)

HABITAT ~ Steep serpentine cliffs near streams or seeps

TYPE AND PLACEMENT ~ Terrestrial

CONSERVATION STATUS ~ Critically endangered due to poaching

FLOWERING TIME ~ April to May (spring)

Often referred to as the "king of orchids," this impressive multi-floral species was named for Ferdinand James von Rothschild (1839–98), a member of the Rothschild banking family and a supporter of horticultural science. Its large size and strong colors have made it a coveted collector's item and an outstanding parent of hybrids. Known from only a few sites on Mount Kinabalu, it has been close to extinction several times due to overzealous collectors.

The outstretched petals have an array of fine hairs and spots that lure flies to these flowers. They try to lay their eggs on the staminode, a sterile stamen, but instead fall into the traplike pouch and pick up a mass of pollen as they exit through the top part of the lip.

The flower of Rothschild's Slipper Orchid has cream-colored sepals and petals, overlaid with bold mahogany stripes and spots. The color of the pouch-shaped, forward-jutting lip varies from light reddish-brown to deep maroon red.

DISTRIBUTION

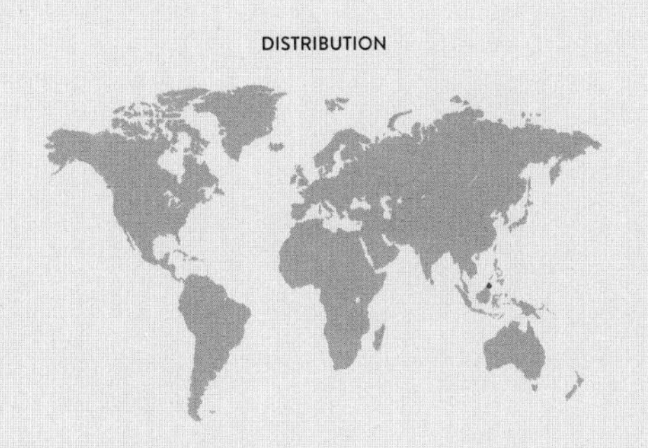

PONTHIEVA MACULATA

SPOTTED SHADOW WITCH

LINDLEY, 1845

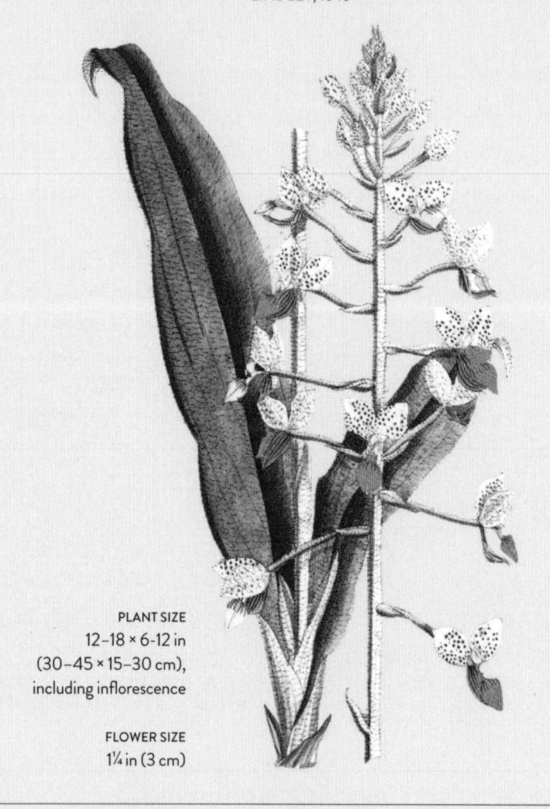

PLANT SIZE
12–18 × 6-12 in
(30–45 × 15–30 cm),
including inflorescence

FLOWER SIZE
1¼ in (3 cm)

SUBFAMILY ~ Orchidoideae

TRIBE AND SUBTRIBE ~ Cranichideae, Cranichidinae

NATIVE RANGE ~ Northwestern South America, Galapagos Islands

HABITAT ~ Rain and cloud forests, at 1,640–9,850 ft (500–3,000 m)

TYPE AND PLACEMENT ~ Terrestrial

CONSERVATION STATUS ~ Not assessed

FLOWERING TIME ~ February to April (late winter to spring)

Species of the genus *Ponthieva* prefer moist sites on the floor of cloud and other wet forests. One species, *P. racemosa*, reaches as far north as Virginia in eastern North America, which is unusual for a principally tropical group of orchids. The 15–20 long-stalked flowers of the Spotted Shadow Witch are carried on a small-bracted hairy inflorescence that emerges from a rosette of hairy, almost stemless leaves with a cluster of fat hairy roots. Oil, rather than nectar, is produced on glands on the lip and lateral sepals, so it is assumed (though not observed) that the pollinators are oil-gathering anthophorid bees.

The flowers are non-resupinate, and, unusually among orchids, it is the erect lateral sepals that are the most attractive part of these flowers. In Costa Rica, the roots are reportedly used as a substitute for ipecacuanha, an emetic.

The flower of the Spotted Shadow Witch is covered with hairs on the outside. The two heavily spotted lateral sepals point upward. The two yellow petals are clawed, with a bulging gland at their base. The lip has a cavity in its center.

DISTRIBUTION

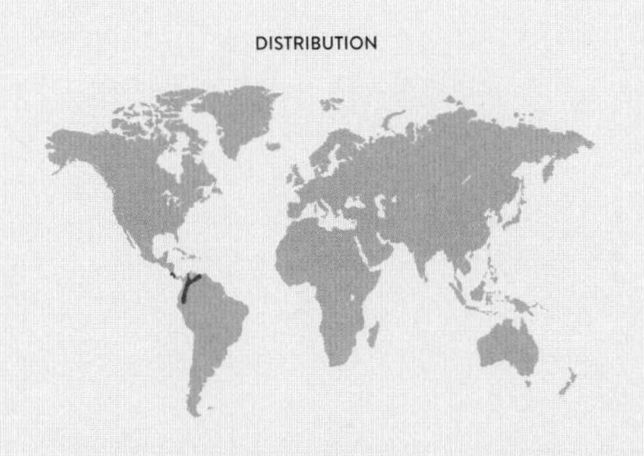

SKEPTROSTACHYS ARECHAVALETANII

SCEPTER ORCHID

(BARBOSA RODRIGUES) GARAY, 1980

PLANT SIZE
14–26 × 6–10 in
(36–66 × 15–25 cm),
including inflorescence

FLOWER SIZE
¼ in (0.75 cm)

SUBFAMILY ~ Orchidoideae

TRIBE AND SUBTRIBE ~ Cranichideae, Spiranthinae

NATIVE RANGE ~ Southern Brazil and Uruguay

HABITAT ~ Open grassland and scrubland, from sea level to 5,250 ft (1,600 m)

TYPE AND PLACEMENT ~ Terrestrial

CONSERVATION STATUS ~ Not assessed

FLOWERING TIME ~ October to December (late spring to early summer)

The Scepter Orchid grows in open areas, including wet meadows and marshes that are subject to seasonal fires. It also occurs in rocky fields, known as *campos rupestres* in Brazil. From a short rhizome, with a cluster of fingerlike, fleshy hairy roots, a central stem is formed with a series of spirally arranged leaves that quickly become smaller and merge with the large pale bracts subtending each flower. The flower stem is topped with 25–50 densely packed flowers arranged in a spike.

There is no information available on pollination of this species. Its similarity in shape and color to *Sacoila lanceolata*, however, might suggest that it is pollinated by hummingbirds.

The flower of the Scepter Orchid is coral to orange red. It is small, fleshy, and does not fully open. The sepals and petals are held forward, and the lip points down between the two lateral sepals, forming a channel into the nectar cavity at the lip base.

DISTRIBUTION

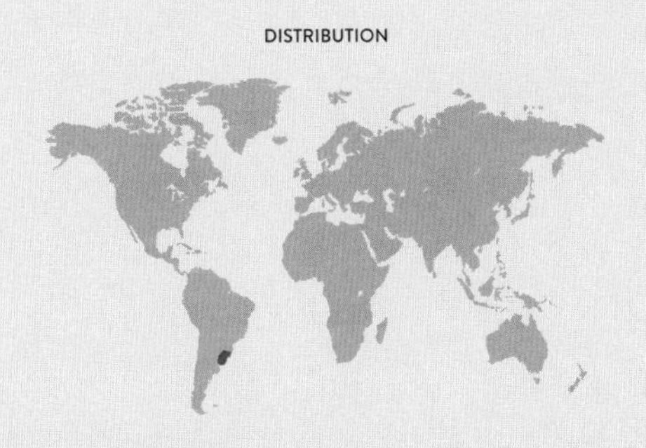

STENORRHYNCHOS SPECIOSUM

VERMILION LADY'S TRESSES

(JACQUIN) RICHARD, 1817

PLANT SIZE
8–36 × 6–10 in
(20–91 × 15–25 cm),
including inflorescence

FLOWER SIZE
¾ in (1.8 cm)

SUBFAMILY ~ Orchidoideae

TRIBE AND SUBTRIBE ~ Cranichideae, Spiranthinae

NATIVE RANGE ~ Mexico, the Caribbean to Peru

HABITAT ~ Moister areas of seasonally dry semi-deciduous forests, often on steep embankments near seeps, at 3,950–9,850 ft (1,200–3,000 m)

TYPE AND PLACEMENT ~ Terrestrial, occasionally epiphytic

CONSERVATION STATUS ~ Not threatened

FLOWERING TIME ~ Winter

One of the more spectacular of the terrestrial orchids, also amenable to cultivation, the widespread Vermilion Lady's Tresses has a basal rosette of spirally arranged, variegated (striped or spotted) leaves reminiscent of the garden plant *Hosta*. True spectacle ensues when brilliant red spikes emerge from the center of the rosettes, bearing dazzling, torchlike racemes of up to 50 (usually 20–30) small red and white flowers, each subtended by a bright red bract. Pollinating hummingbirds are irresistibly drawn to these flowers, which are waxy and tough enough to stand up to the onslaught of a bird's beak.

The handsome plants have a cluster of thick, hairy roots that sustain them through dry seasons when leaves wither. This dormancy usually occurs shortly after blooming takes place. Some forms of this species have entirely green leaves without any variegation.

The flower of the Vermilion Lady's Tresses is small, cupped, and white but infused with brilliant red and subtended by a dazzling red bract, which makes the blooms and inflorescence appear to be completely red.

DISTRIBUTION

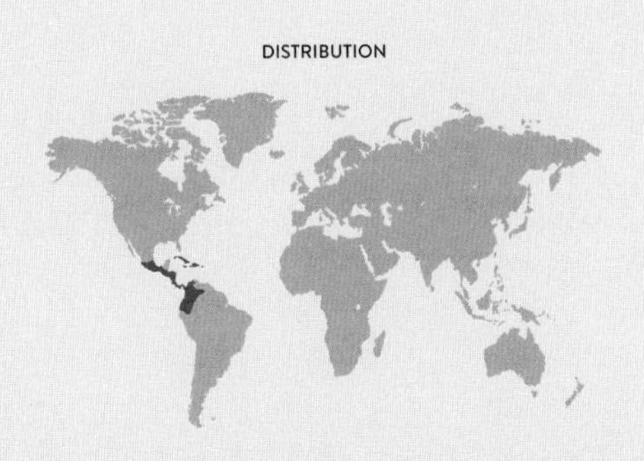

CORYBAS PICTUS

PAINTED HELMET ORCHID

(BLUME) REICHENBACH FILS, 1871

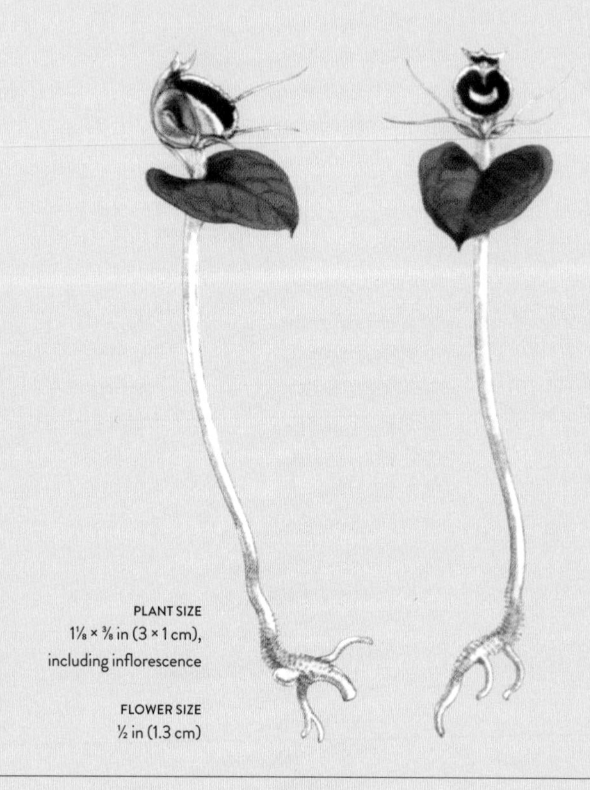

PLANT SIZE
1⅛ × ⅜ in (3 × 1 cm), including inflorescence

FLOWER SIZE
½ in (1.3 cm)

SUBFAMILY ~ Orchidoideae

TRIBE AND SUBTRIBE ~ Diurideae, Acianthinae

NATIVE RANGE ~ Malaysia and western Indonesia (Java, Sumatra, Borneo)

HABITAT ~ Middle elevation forests

TYPE AND PLACEMENT ~ Terrestrial on mossy rocks and tree trunks, usually on steep slopes

CONSERVATION STATUS ~ Not assessed but rare

FLOWERING TIME ~ October to February

This delicate, tiny orchid looks almost extraterrestrial with its hood and long tentacles. It is a tropical species that lives in mountain forests, where it needs dark and moist conditions and often grows on fibrous trunks of tree ferns or on mossy trunks of other forest trees. The name *pictus* derives from the Latin *pingere*, meaning "to decorate," referring to the attractive white venation on the tiny leaf.

Like most species of *Corybas*, the Painted Helmet Orchid is pollinated by fungus gnats. The long extensions on the sepals and petals occur only in the tropical species of this genus, and it is thought this may help lead flies to the center of the flower. The function of such floral specializations has never been experimentally tested, and such statements about potential roles are only speculative.

The flower of the Painted Helmet Orchid sits atop a heart-shaped, silver-veined leaf. The upper sepal (helmet) is dark brownish-red with a white edge. The lip has a white "cushion," and sepals and petals are elongated into antennae.

DISTRIBUTION

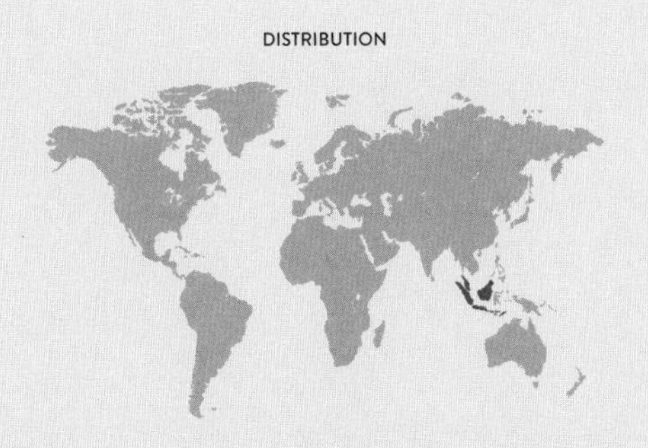

CALADENIA LONGICAUDA

LARGE WHITE SPIDER ORCHID

LINDLEY, 1839

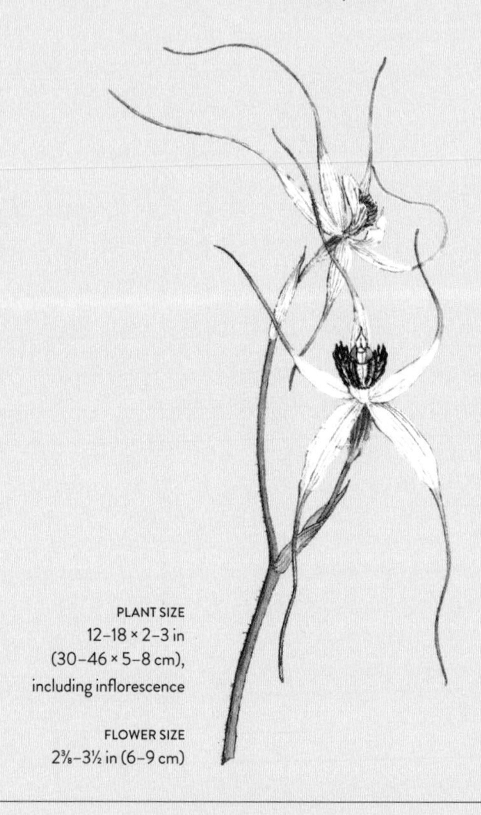

PLANT SIZE
12–18 × 2–3 in
(30–46 × 5–8 cm),
including inflorescence

FLOWER SIZE
2⅜–3½ in (6–9 cm)

SUBFAMILY ~ Orchidoideae

TRIBE AND SUBTRIBE ~ Diurideae, Caladeniinae

NATIVE RANGE ~ Southwestern Australia

HABITAT ~ Forests and rocky outcrops, swamps, lake margins, and coastal scrub, at up to 985 ft (300 m)

TYPE AND PLACEMENT ~ Terrestrial on sand, clayey loam, laterite, or gravel

CONSERVATION STATUS ~ Not threatened

FLOWERING TIME ~ July to November (winter to spring)

Also called "daddy long legs," the Large White Spider Orchid, found close to Perth, has beautiful spidery flowers and is emblematic of the spring flora of Western Australia. It has an underground tuber with which it survives the dry season and fires. A single leaf and an inflorescence of one or six large flowers are produced, and both leaf and stem are hairy.

Spider orchid species are most likely pollinated by food deceit, and visits by a large variety of bees, wasps, and flies, all species that feed on nectar and pollen, have been reported. Some authors have suggested that these white-flowered species exhibit a mixed pollination syndrome and are also sexually attractive to male thynnine wasps, but this has yet to be demonstrated.

The flower of the Large White Spider Orchid has long and spidery sepals and petals. Four are reflexed, but the dorsal sepal is held upright or over the column. The lip is white, recurving, and fringed with brown to purple-red hairs.

DISTRIBUTION

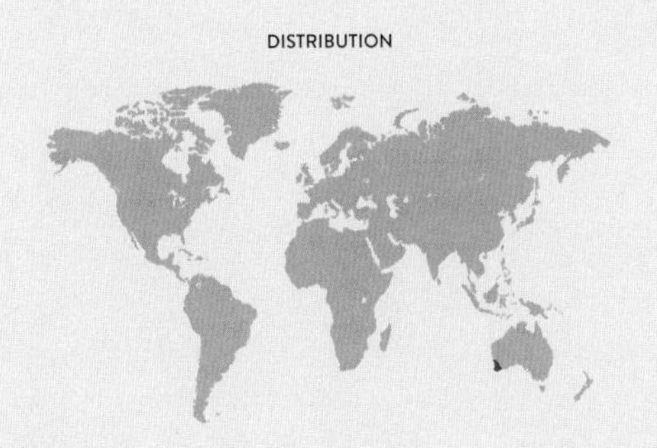

DISA GRAMINIFOLIA

BLUE MOTHER'S CAP

KER GAWLER EX SPRENGEL, 1826

PLANT SIZE
40 × 10 in (102 × 25 cm), including inflorescence

FLOWER SIZE
1½ in (3.8 cm)

SUBFAMILY ~ Orchidoideae

TRIBE AND SUBTRIBE ~ Orchideae, Disinae

NATIVE RANGE ~ Southwestern and southern Cape province of South Africa,

HABITAT ~ Dense scrub on dry sandstone in full sun, at 985–3,300 ft (300–1,000 m)

TYPE AND PLACEMENT ~ Terrestrial

CONSERVATION STATUS ~ Not assessed, but locally common

FLOWERING TIME ~ January to March (summer)

In Afrikaans, the orchid name is *bloumoederkappie*, meaning "blue mother's cap," which refers to the similarity of the flower to a traditional female headdress. The species bears up to ten magnificent blue flowers per inflorescence after the grasslike leaves have died off during the summer. Underground, there is a globose tuber, which is eaten locally and used to make flour for a type of bread.

The flowers have a sweet scent, but although they offer no nectar, carpenter bees frequent them in search of new nectar sources. The bees enter by pushing the robust column aside, and then pollinia become glued to their thorax. The floral differences from more typical species of the genus *Disa* resulted in *D. graminifolia* being widely known as *Herschelianthe graminifolia*, but most botanists today treat it as part of *Disa*.

The flower of Blue Mother's Cap has highly ornamented petals, often with purple and green spots, which flank the column under the hood. One sepal forms a club-shaped, hooded, upward-pointing spur. The lip is dark purple with downcurved margins, fading to white in the middle.

DISTRIBUTION

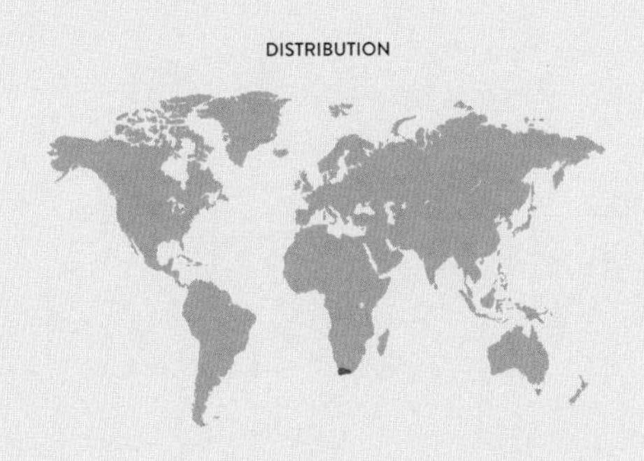

BONATEA SPECIOSA

BEAUTIFUL REIN ORCHID

(LINNAEUS FILS) WILLDENOW, 1805

PLANT SIZE
15–45 × 8–20 in
(38–114 × 20–51 cm)

FLOWER SIZE
2–2⅜ in (5–6 cm)

SUBFAMILY ~ Orchidoideae

TRIBE AND SUBTRIBE ~ Orchideae, Orchidinae

NATIVE RANGE ~ Coastal Mozambique to southern South Africa

HABITAT ~ Coastal bush, open deciduous woodlands, and forest edges at low elevation

TYPE AND PLACEMENT ~ Terrestrial

CONSERVATION STATUS ~ Not threatened

FLOWERING TIME ~ Spring and summer

As its common name suggests, the stately *Bonatea speciosa* has beautiful flowers, densely arranged on an upright raceme. Found in sandy, well-drained soils, the plant dies back to its large, woolly cylindrical tubers during the winter dry season. It is one of the most distinctive and plentiful South African orchid species, well known for more than 200 years, and, unlike many terrestrial orchids, is easily and commonly cultivated for its unusual, hooded blooms.

The plant is pollinated by a hawk moth, attracted to its stunning green and white, nocturnally fragrant spurred blossoms. This odd structure has been the subject of controversy since its first description. Thought originally to have a five-lobed lip, it instead has petals with two lobes, one of which fuses with the labellum to give the appearance of two additional lobes.

The flower of the Beautiful Rein Orchid is green and white with a hooded dorsal sepal. The bizarrely two-lobed petals form what appear to be more labellum lobes.

DISTRIBUTION

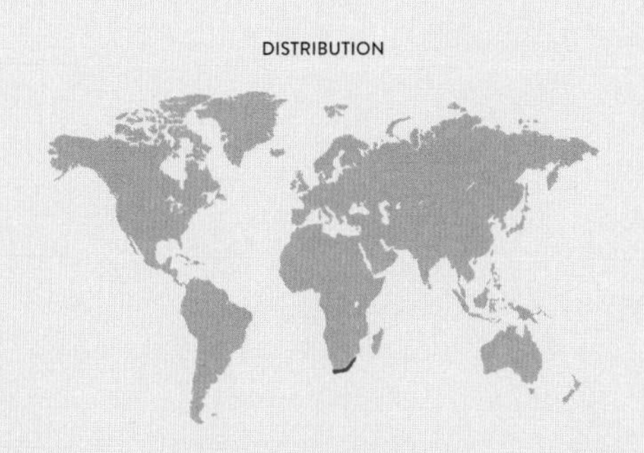

ARUNDINA GRAMINIFOLIA

BAMBOO ORCHID

(D. DON) HOCHREUTINER, 1910

PLANT SIZE
50–100 × 12–30 in
(127–254 × 31–76 cm),
including inflorescence, which
occurs at the tips of the tall plants

FLOWER SIZE
2½ in (6.5 cm)

SUBFAMILY ~ Epidendroideae

TRIBE AND SUBTRIBE ~ Arethuseae, Arethusinae

NATIVE RANGE ~ Tropical and subtropical Asia; also naturalized in the American and African tropics

HABITAT ~ Exposed situations, rocks, lava flows, and in meadows, up to 3,950 ft (1,200 m)

TYPE AND PLACEMENT ~ Terrestrial

CONSERVATION STATUS ~ A common species in its native and non-native areas

FLOWERING TIME ~ January to December

The Bamboo Orchid is one of the most commonly encountered through the tropical regions of the world. It has escaped from gardens into the wild on many occasions, often becoming part of island floras, in places such as Puerto Rico, Jamaica, Guadeloupe, Hawaii, and Réunion, where it colonizes fresh lava flows. Alternating lanceolate leaves produced by a tall stem make the plant look like a bamboo. Each stem bears a bract-bearing inflorescence that carries up to six fragrant flowers, produced one at a time.

The flowers lack nectar but have extrafloral nectaries on the inflorescence, which are visited by many insects, especially ants. As its fruit set is often good, the species is probably self-pollinating. Otherwise, in many places where the orchid grows naturally or has been introduced, pollination is by bees of the genus *Xylocopa*.

The flower of the Bamboo Orchid is usually white to pale lavender with a bright purple labellum. It has narrow sepals—one upright, the others folded behind the trumpet-shaped lip that envelops the column, which often has a yellow, central spot. Petals are spreading, broad, and showy.

DISTRIBUTION

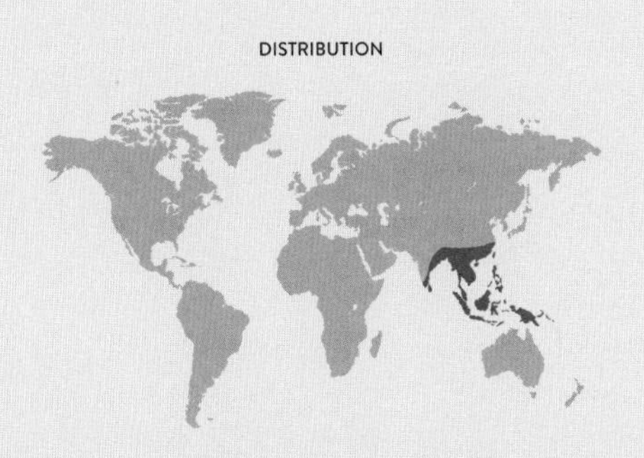

BLETILLA STRIATA

ASIAN HYACINTH ORCHID

(THUNBERG) REICHENBACH FILS, 1878

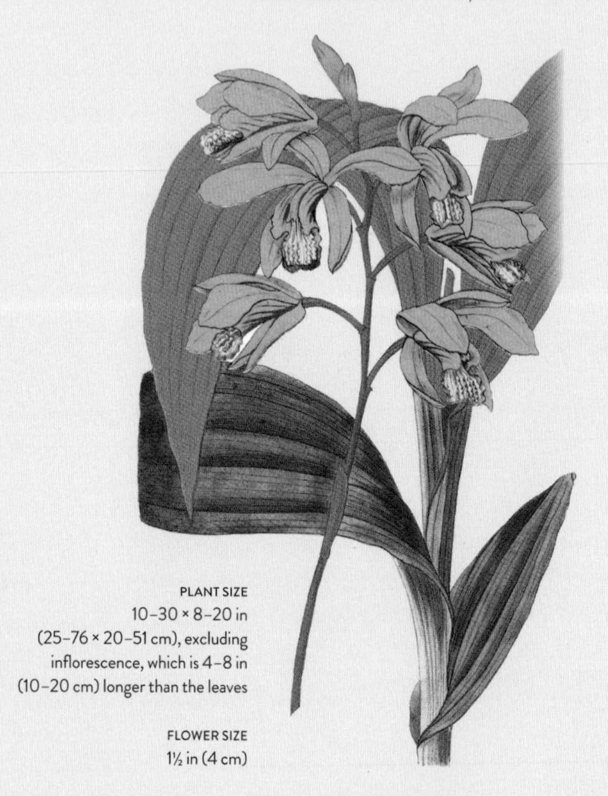

PLANT SIZE
10–30 × 8–20 in
(25–76 × 20–51 cm), excluding inflorescence, which is 4–8 in (10–20 cm) longer than the leaves

FLOWER SIZE
1½ in (4 cm)

SUBFAMILY ~ Epidendroideae

TRIBE AND SUBTRIBE ~ Arethuseae, Coelogyninae

NATIVE RANGE ~ Northern Myanmar, throughout China and Korea to Japan

HABITAT ~ Evergreen broadleaved or coniferous forests, grassy meadows, or rock crevices, at 330–10,500 ft (100–3,200 m)

TYPE AND PLACEMENT ~ Terrestrial

CONSERVATION STATUS ~ Endangered in the wild, due to overcollecting, but commonly and easily cultivated

FLOWERING TIME ~ April to June (spring)

Large plicate leaves grow from the corm-like pseudobulbs of the Asian Hyacinth Orchid in spring. Among these leaves, a raceme emerges bearing a few, fragrant, showy, nodding flowers. These are usually pinkish-purple, though forms with white flowers or variegated leaves are known in horticulture. In Japan, pollination has been reported to be by male and female longhorn bees, *Tetralonia nipponensis*.

In Chinese medicine, the astringent bitter pseudobulbs of *Bletilla* are commonly used in a mix of medicinal compounds to reduce swelling and stop bleeding associated with lung, stomach, and liver disorders, and to promote tissue healing. The pseudobulbs also produce a mucilage, used in the production of ceramics. *Bletilla* root is sold in many Asian markets and specialty shops, but can also be cultivated, as the plant is hardy and easy to grow.

The flower of the Asian Hyacinth Orchid has dark pink sepals and petals, which are slightly lobed. The three-lobed lip forms a tube and is hinged to the column base, the middle lip having lamellae.

DISTRIBUTION

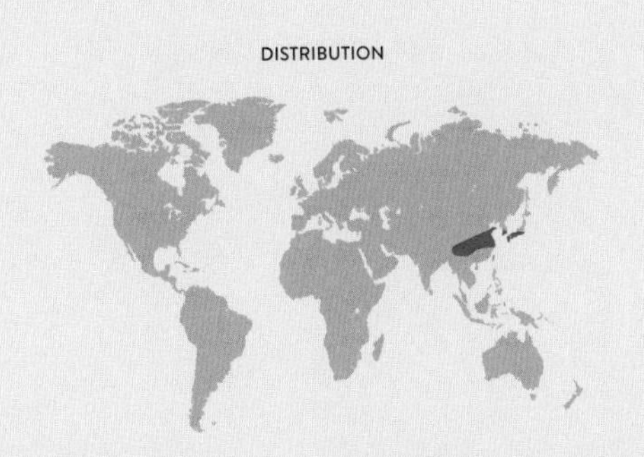

COELOGYNE PANDURATA

BLACK FIDDLE ORCHID

LINDLEY, 1853

PLANT SIZE
15–28 × 8–12 in (38–71 × 20–30 cm), excluding arching to pendent 10–30 in (25–76 cm) long inflorescence, which can be longer than the leaves

FLOWER SIZE
3 in (8 cm)

SUBFAMILY ~ Epidendroideae

TRIBE AND SUBTRIBE ~ Arethuseae, Coelogyninae

NATIVE RANGE ~ Malaysia, Sumatra, Borneo, and the Philippines

HABITAT ~ Tropical wet forests, often near streams, at 4,950–6,600 ft (1,500–2,000 m)

TYPE AND PLACEMENT ~ Epiphytic, climber

CONSERVATION STATUS ~ Not threatened

FLOWERING TIME ~ Mostly in summer

The Black Fiddle Orchid is renowned for its unusual black lip, shaped like a fiddle or lute, hence both the common name and scientific name, from the Latin for "fiddle-like." The large flattened pseudobulbs are widely spaced on a stem that climbs around large tree trunks. The impressive blooms, up to 15 at a time, open all at once and smell strongly of honey, but last for only a few days.

Although *Coelogyne pandurata* blooms almost year-round, many of the other species in the genus come from seasonally dry habitats where their pseudobulbs often shrivel alarmingly just prior to blooming. Pollination of these fantastically beautiful flowers has not been observed in nature, but the lack of a spur and the sidelobes of lip surrounding the column would most likely be adapted for a bee or wasp. There is no reward offered, in spite of the honey scent.

The flower of the Black Fiddle Orchid has chartreuse sepals and petals. The large, greenish lip is overlaid with black spots and stripes ornamented with a series of ridges, knobs, and keels, and its margin is frilly.

DISTRIBUTION

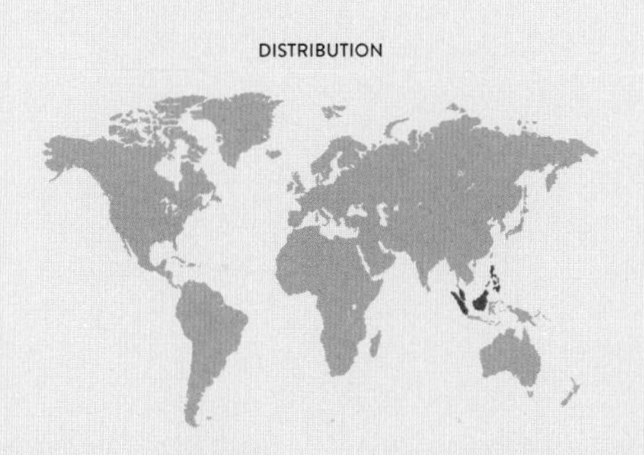

PLEIONE FORMOSANA

FORMOSAN ROCK ORCHID

HAYATA, 1911

PLANT SIZE
6–12 × 3–5 in (15–30 × 8–13 cm), including mostly single-flowered, erect inflorescence, which at the time of flowering is leafless

FLOWER SIZE
3 in (8 cm)

SUBFAMILY ~ Epidendroideae

TRIBE AND SUBTRIBE ~ Arethuseae, Coelogyninae

NATIVE RANGE ~ Southeastern China and Taiwan

HABITAT ~ Seasonally dry, cold forests, at 4,920–8,200 ft (1,500–2,500 m)

TYPE AND PLACEMENT ~ Lithophytic, terrestrial in moss, epiphytic on tree bases

CONSERVATION STATUS ~ Not threatened

FLOWERING TIME ~ Spring

The beautiful Formosan Rock Orchid bears showy blooms that emerge from leafless, cone-shaped, ridged pseudobulbs. Plants occur near the frost zone and often grow on vertical cliff faces, moss-encrusted rocks, and around bases of trees. A new pseudobulb and leaf appear on a newly emerging growth after the separate inflorescence, produced by the pseudobulb from the previous year, has withered. The genus name is the Greek word for "annual" —a reference to the annual cycle of leaf production and loss.

Delightfully fragrant and displaying attractive nectar guides on its lip, *Pleione formosana* attracts bumblebees (*Bombus eximius*, *B. flavescens*, and *B. trifasciatus*) as its pollinators, but there is no nectar reward for the attention of both queen and worker bees. This and other species in the genus are used in Chinese traditional medicine to treat tumors.

The flower of the Formosan Rock Orchid usually comes in shades of lavender purple with lanceolate sepals and petals that are generally similar in shape and color. The heavily fringed lip has reddish markings and yellow keels in the throat, and its sides wrap around the column, forming a tube.

DISTRIBUTION

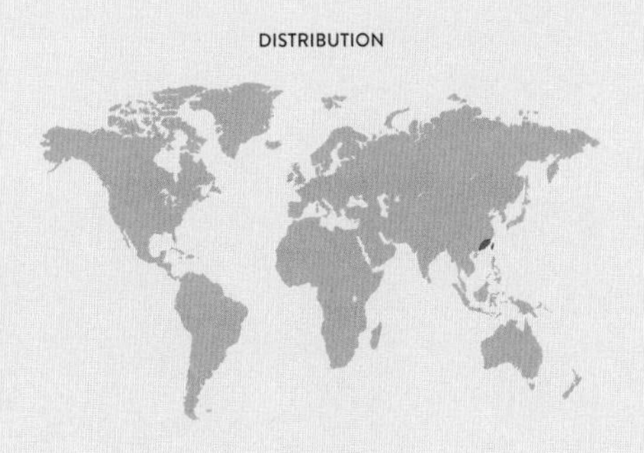

THUNIA ALBA

WHITE BAMBOO ORCHID

(LINDLEY) REICHENBACH FILS, 1852

PLANT SIZE
24–40 × 12–20 in (61–102 × 30–51 cm), including nodding terminal inflorescence 8–12 in (20–30 cm) long

FLOWER SIZE
3½ in (9 cm)

SUBFAMILY ~ Epidendroideae

TRIBE AND SUBTRIBE ~ Arethuseae, Coelogyninae

NATIVE RANGE ~ Southeast Asia, from the Himalayas to southern China and Peninsular Malaysia

HABITAT ~ Forests or shaded rocky places, at 3,300–7,545 ft (1,000–2,300 m)

TYPE AND PLACEMENT ~ Epiphytic on lower tree branches or lithophytic on rocks

CONSERVATION STATUS ~ Not assessed, but due to its broad distribution not likely to be of conservation concern

FLOWERING TIME ~ June (summer)

The White Bamboo Orchid is a big plant, with upright to pendent stems (in sunny as opposed to shady sites) often bearing dozens of winter deciduous leaves on meter-long canes—like a bamboo in habit, hence the common name. It produces large, strongly scented, short-lived flowers.

Pollination has not been observed in *Thunia alba*, but the flower shape, with the lip surrounding the column, and the presence of a spur and nectar guides suggest that bees are the most likely visitors. There are two color forms, one with yellow in the throat of the lip and the other with darker veins covering the series of ridges on the lip.

The flower of the White Bamboo Orchid has spreading, white, lanceolate sepals and petals. The lip surrounds the column, is hairy inside, and bears a yellow blotch, which in some color forms may have darker orange to purple-red veins.

DISTRIBUTION

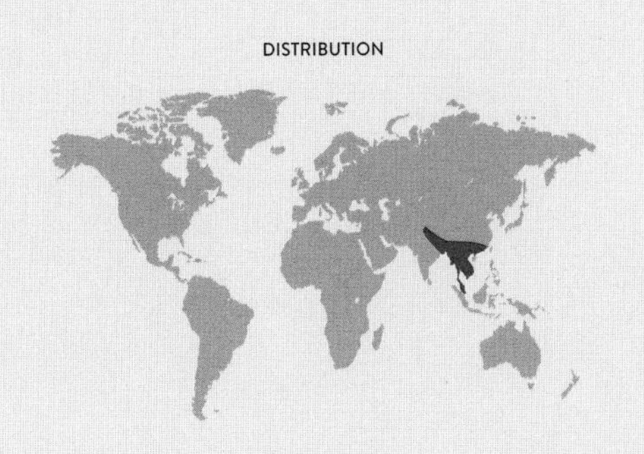

CALANTHE SYLVATICA

BROAD-LEAVED FOREST ORCHID

(THOUARS) LINDLEY, 1833

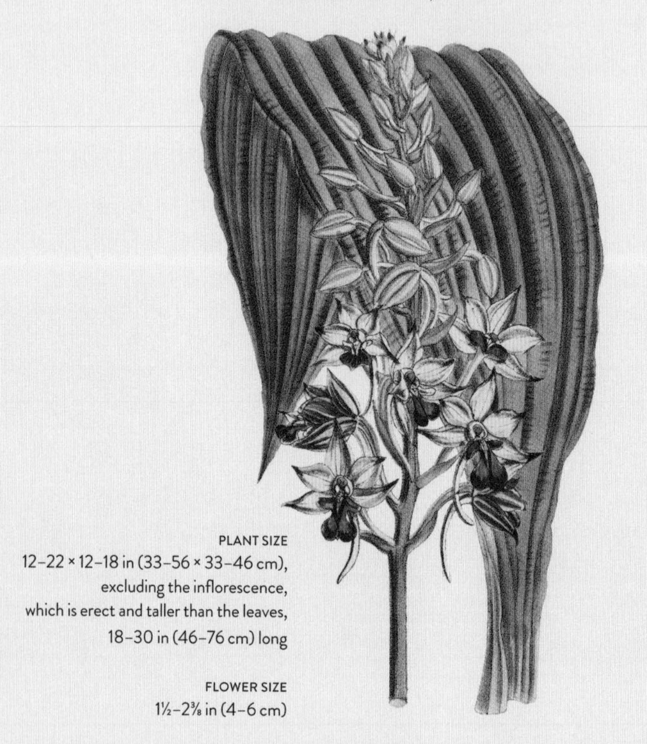

PLANT SIZE
12–22 × 12–18 in (33–56 × 33–46 cm), excluding the inflorescence, which is erect and taller than the leaves, 18–30 in (46–76 cm) long

FLOWER SIZE
1½–2⅜ in (4–6 cm)

SUBFAMILY ~ Epidendroideae

TRIBE ~ Collabieae

NATIVE RANGE ~ Tropical and southern Africa, from Tanzania to South Africa, Madagascar, and Mascarene Islands

HABITAT ~ Stream sides or wet, shady forests

TYPE AND PLACEMENT ~ Terrestrial

CONSERVATION STATUS ~ Not threatened

FLOWERING TIME ~ Late winter to early spring

With a name derived from the Greek words for "beautiful flower" (*calos*, meaning "beautiful," and *anthos*, "flower") *Calanthe* is a genus of mostly terrestrial orchids, widely admired for their stately spikes of colorful blooms, set against striking broad, plicate foliage typical of shade-loving forest understory plants. Possibly one of the earliest tropical orchids planted in a garden setting, the very first orchid hybrid, *C. dominyi*, was bred from this lovely species.

The colorful well-spaced blossoms grow from sturdy upright spikes up to 16 in (40 cm) long and bloom over a long period. Usually pale purple or mauve, the flowers can display richer, vibrant purples and other colors, and often take on an orangey hue as they fade. Although many related species are deciduous in temperate zones, this tropical member of the genus is evergreen and not able to take a freeze.

The flower of the Broad-leaved Forest Orchid is deep or pale purple to violet and white and held on tall spikes above the foliage. Sepals are falcate or winglike, whereas the petals are narrower. The lip is trilobed with an indented midlobe and bears a central yellow to orange crest.

DISTRIBUTION

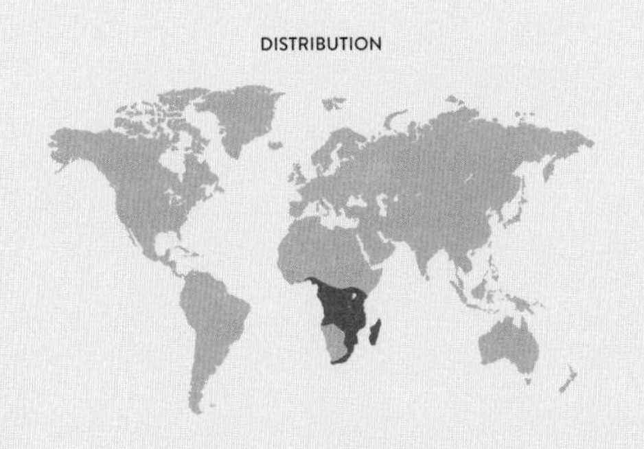

PHAIUS TANKERVILLEAE

GREATER SWAMP ORCHID

(BANKS) BLUME, 1856

PLANT SIZE
35–60 × 20–35 in (89–152 × 51–89 cm), excluding erect inflorescence 40–70 in (102–178 cm) tall, which is longer than the leaves

FLOWER SIZE
4¾ in (12 cm)

SUBFAMILY ~ Epidendroideae

TRIBE ~ Collabieae

NATIVE RANGE ~ Southern China, Southeast Asia, and northern Australia to the Pacific islands; escaped from cultivation and becoming invasive in Hawaii, Florida, and other suitably tropical areas

HABITAT ~ Shady to bright wet-forest margins, from sea level to 2,625 ft (800 m)

TYPE AND PLACEMENT ~ Terrestrial

CONSERVATION STATUS ~ Not threatened

FLOWERING TIME ~ Late spring

The striking Greater Swamp Orchid is popular in horticulture, both under glass and outside in tropical regions, where it has escaped repeatedly into nature. The genus name derives from the Greek word *phios*, "gray," referring to the peculiar gray color of the flowers as they die. This is due to the presence of indigo, which in the past was extracted and used in New Guinea to dye clothing. The plants are also employed in folk medicine, as a poultice for infected sores in Java and as an aid to conception in New Guinea. The species name honors Lady Emma Colebrooke, Countess of Tankerville (1752–1836), in whose collection at Walton-on-Thames, near London, it first flowered.

The stately upright stems hold 10–35 flowers. Pollination is by carpenter bees (*Xylocopa*), although no reward is offered. After flowering, the stems produce plantlets, which permits them to form large colonies.

The flower of the Greater Swamp Orchid has tan to pinkish-tan lanceolate sepals and petals with a white or pale yellowish reverse. The lip is pink to purple, basally fading to near white, and the sidelobes wrap around the column. White and yellow forms occur.

DISTRIBUTION

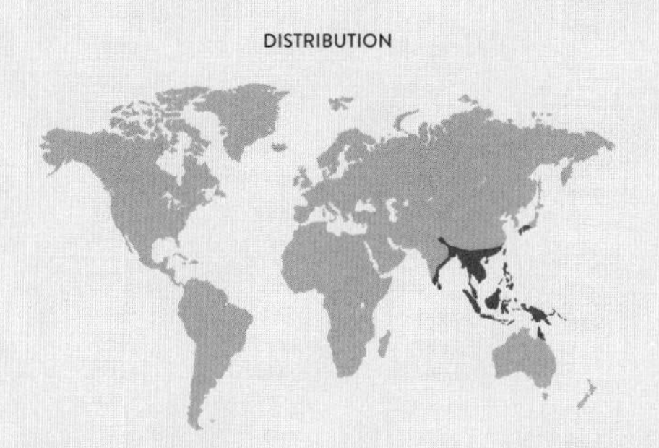

CLOWESIA ROSEA

ROSE-COLORED BASKET ORCHID

LINDLEY, 1843

PLANT SIZE
10–18 in (25–46 cm), excluding pendent inflorescence 4–8 in (10–20 cm) long

FLOWER SIZE
1½–2⅜ in (4–6 cm)

SUBFAMILY ~ Epidendroideae

TRIBE AND SUBTRIBE ~ Cymbidieae, Catasetinae

NATIVE RANGE ~ Southern Mexico to Guatemala

HABITAT ~ Seasonally dry deciduous forests, at 1,650–4,950 ft (500–1,500 m)

TYPE AND PLACEMENT ~ Epiphytic

CONSERVATION STATUS ~ Not threatened

FLOWERING TIME ~ Late winter to early spring

The seasonally dry Pacific slopes of southern Mexico are home to some extraordinary species, including the Rose-colored Basket Orchid, which, with its pendent racemes of attractive flowers, has been a favorite of orchid connoisseurs for over a century. The flowers emit a strong, delicious fragrance of cinnamon and citrus that is collected by male euglossine bees and used to produce a cologne attractive to discriminating females. Plants of genus *Clowesia* differ from the closely related genus *Catasetum* by having bisexual flowers rather than separate male or female flowers with different shapes.

The roots of the species first grow downward and then branch, the branches growing upward to form a leaf litter-trapping basket—giving this species its common name. This trait is rare in orchids and known to occur only in a small number of distantly related genera.

The flower of the Rose-colored Basket Orchid is pale pink, overlaid on white or cream segments. The petals are fringed on their margin, and the gullet-shaped lip is deeply fringed on its margin. The crest and column are pale yellow.

DISTRIBUTION

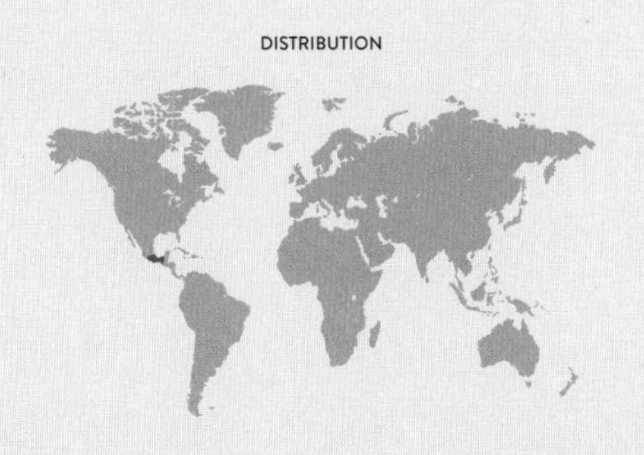

MORMODES COLOSSUS

FLYING BIRD ORCHID

REICHENBACH FILS, 1852

PLANT SIZE
24–32 × 15–24 in (61–81 × 38–61 cm) excluding the erect inflorescence, 18–28 in (46–71 cm) tall

FLOWER SIZE
Up to 4¾ in (12 cm)

SUBFAMILY ~ Epidendroideae

TRIBE AND SUBTRIBE ~ Cymbidieae, Catasetinae

NATIVE RANGE ~ Costa Rica and Panama

HABITAT ~ Cloud forests, at 3,950–6,900 ft (1,200–2,100 m)

TYPE AND PLACEMENT ~ Epiphytic

CONSERVATION STATUS ~ Has a restricted distribution but occurs in reserves, therefore unlikely to be under threat of extinction

FLOWERING TIME ~ Spring

The large Flying Bird Orchid has cylindrical, tapering pseudobulbs with several (usually two to five) plicate leaves, which are deciduous in the dry season. At the base of the pseudobulbs, a long arching raceme emerges carrying 6–15 fragrant flowers that vary remarkably in size and color. The genus name comes from *Mormo*, a malevolent female spirit in Greek mythology, and *–oides*, "resembling," referring to the strange appearance of the flowers. The plants are often found on rotting stumps or fallen branches and have a mycorrhizal connection to wood-rot fungi.

Flowers of *Mormodes* are functionally male until the pollinia are removed. A long slender appendage on the anther cap is brushed by visiting male euglossine bees, causing the pollinarium to become attached to their back. The coiled stipe of the pollinarium takes 30 minutes to straighten and assume the correct position for pollination.

The flower of the Flying Bird Orchid has spreading to recurved, lanceolate sepals and petals. The lip is shaped like a boat, with one end stalked and fixed to the flower, the other end pointing upward, overarching the column.

DISTRIBUTION

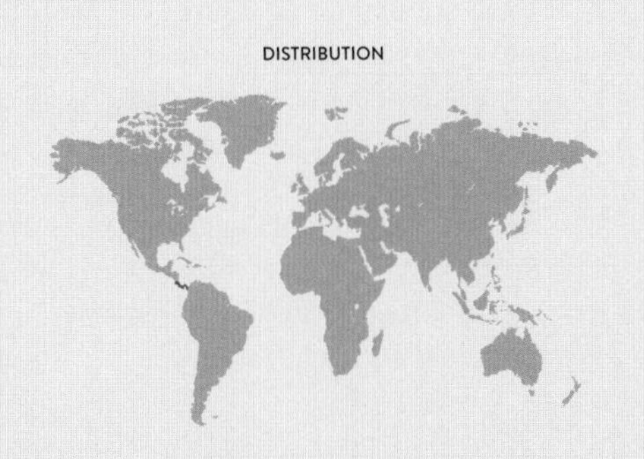

GRAMMATOPHYLLUM SPECIOSUM

SHOWY TIGER ORCHID

BLUME, 1825

PLANT SIZE
40–400 × 20–38 in (102–1,016 × 51–97 cm), excluding erect-arching inflorescence 100–500 in (254–1270 cm) tall, which is taller than the leafy stems

FLOWER SIZE
5 in (12.5 cm)

SUBFAMILY ~ Epidendroideae

TRIBE AND SUBTRIBE ~ Cymbidieae, Cymbidiinae

NATIVE RANGE ~ Myanmar, Thailand, Laos, Vietnam, Malaysia, and Indonesia

HABITAT ~ Hot wet forests, near streams and rivers, at 330–2,625 ft (100–800 m)

TYPE AND PLACEMENT ~ Epiphytic, occasional lithophytic

CONSERVATION STATUS ~ Threatened by poaching

FLOWERING TIME ~ July to October

Considered by many to be the world's largest orchid (but not the tallest), the Showy Tiger Orchid is unquestionably the heaviest of all orchids. Its many long, lanceolate leaves borne on impressive cane-like pseudobulbs give the plants, at least superficially, the look of a palm. The genus name comes from the Greek words *gramma*, "letter," and *phyllon*, "leaf," a reference to the bold marking on the sepals and petals. The species name refers to the massive display of flowers, from the Latin for "showy."

Pollination of this species has been reported to be by large carpenter bees (*Xylocopa*) seeking nectar, although the flowers produce none. Low on the inflorescence are several sterile flowers (no column or lip are present), which produce only fragrance.

The flower of the Showy Tiger Orchid generally has large yellow sepals and petals with an overlay of chestnut-brown spots and blotches. The lip is much smaller, usually pale yellow with some red marking on the callus and sidelobes, which surround the column.

DISTRIBUTION

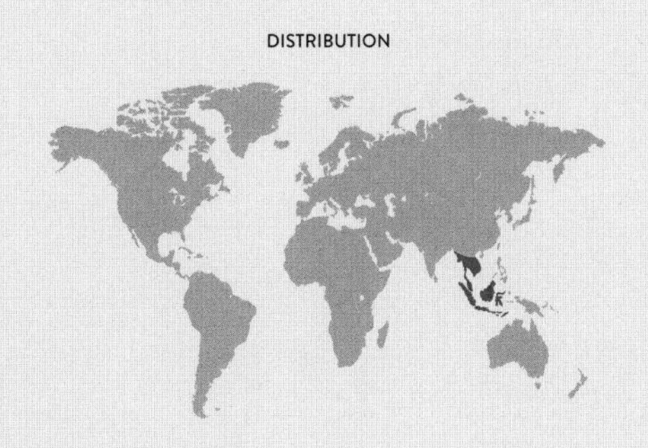

LYCASTE AROMATICA

CINNAMON ORCHID

(GRAHAM) LINDLEY, 1843

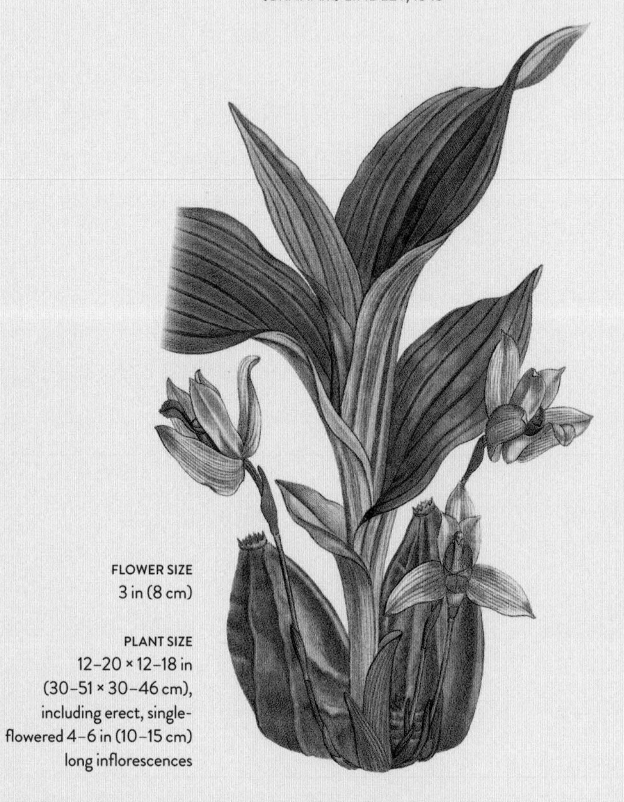

FLOWER SIZE
3 in (8 cm)

PLANT SIZE
12–20 × 12–18 in (30–51 × 30–46 cm), including erect, single-flowered 4–6 in (10–15 cm) long inflorescences

---

SUBFAMILY ~ Epidendroideae

TRIBE AND SUBTRIBE ~ Cymbidieae, Maxillariinae

NATIVE RANGE ~ Mexico, Guatemala, Nicaragua, Belize, Honduras, and El Salvador

HABITAT ~ Seasonally dry forests, at 1,640–6,600 ft (500–2,000 m)

TYPE AND PLACEMENT ~ Epiphytic, terrestrial, and lithophytic

CONSERVATION STATUS ~ Threatened by overcollection

FLOWERING TIME ~ Late spring to summer

The Cinnamon Orchid produces up to ten vivid yellow, deliciously fragrant blossoms on short individual inflorescences from each leafless pseudobulb as the next season's growth is starting to emerge. Extremely adaptable, it can be found growing epiphytically on mossy branches as well as on steep seasonally moist limestone outcrops. Plants are deciduous during the dry season, with the bulbs bearing sharp spines at their apex. The genus name derives from Lycaste, the beautiful daughter of King Priam of Troy.

The sweet cinnamon aroma emitted by the flowers is a lure for male euglossine bees, which use the fragrance compounds to produce a sex pheromone that attracts females. There are several related species that look similar to *Lycaste aromatica*, but these display slightly different blooming seasons and fragrances.

The flower of the Cinnamon Orchid is greenish-yellow with ovate, acuminate sepals arranged in a triangle and brilliant yellow, incurved petals. The also bright yellow, three-lobed lip has a pubescent disc and a large, ridged callus that extends over the midlobe.

DISTRIBUTION

MAXILLARIA TENUIFOLIA

COCONUT ORCHID

LINDLEY, 1837

PLANT SIZE
10–18 × 3–5 in (25–46 × 8–13 cm), including single-flowered, erect 2–4-in (5–10-cm) tall inflorescences

FLOWER SIZE
2 in (5 cm)

SUBFAMILY ~ Epidendroideae

TRIBE AND SUBTRIBE ~ Cymbidieae, Maxillariinae

NATIVE RANGE ~ Mexico, Guatemala, El Salvador, Honduras, Nicaragua, and Costa Rica

HABITAT ~ Wet tropical forests, at 1,970–4,920 ft (600–1,500 m)

TYPE AND PLACEMENT ~ Epiphytic, rarely terrestrial on slopes or embankments

CONSERVATION STATUS ~ Not threatened

FLOWERING TIME ~ Mostly in spring and summer

Due to its delicious coconut fragrance and hardy nature, the Coconut Orchid is widely cultivated. It is a vigorous species, native to a wide range and tolerant of climatic conditions through much of Central America. It forms large clumps on tree trunks and major branches, with many small bulbs bearing one long, narrow apical leaf. The genus name is from the Latin word for "jaw," *maxilla*, a reference to the "chin" formed by the bases of the lip and column. The species name, from the Latin for "slender-leaved," refers to the narrow, grasslike leaves.

The fragrant blooms arise on short stems from the bases of the newest pseudobulbs. *Maxillaria tenuifolia* uses fragrance to lure bees, which in this case are meliponine bees (stingless bees). The pollinaria are attached to the scutellum (forward portion of the thorax), and in this case there is no reward for the pollinator.

The flower of the Coconut Orchid is pale yellow, heavily overlaid with brown to maroon red blotches that coalesce into a solid red color on the tips. The lip is buff yellow with maroon red spotting. The column is creamy white.

DISTRIBUTION

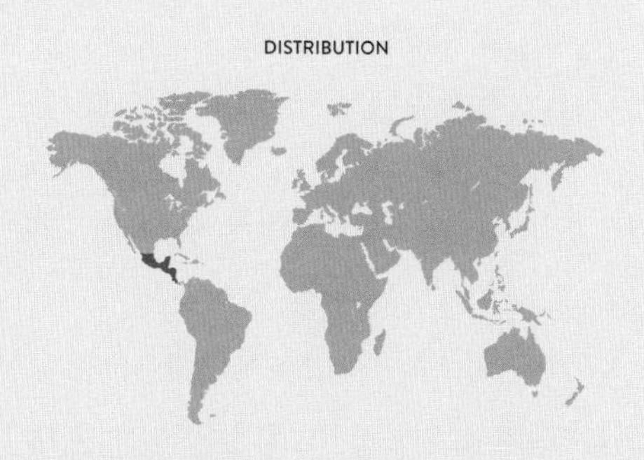

BRASSIA AURANTIACA

ORANGE CLAW

(LINDLEY) M. W. CHASE, 2011

PLANT SIZE
8–14 × 8–12 in
(20–36 × 20–30 cm),
excluding arching inflorescence 12–20 in
(30–51 cm) long, which is longer than the leaves

FLOWER SIZE
1¼ in (3 cm)

SUBFAMILY ~ Epidendroideae

TRIBE AND SUBTRIBE ~ Cymbidieae, Oncidiinae

NATIVE RANGE ~ Venezuela, Colombia, and Ecuador

HABITAT ~ Cold wet tropical forests, at 6,600–8,250 ft (2,000–2,500 m)

TYPE AND PLACEMENT ~ Epiphytic

CONSERVATION STATUS ~ Not threatened

FLOWERING TIME ~ Late winter to early spring

The flowers of the Orange Claw, unlike those of most *Brassia* species, do not open fully and are tightly bunched, with up to 18 flowers per stem. What the plants lack in individual flower form, they more than make up for in their displays of many brightly colored flowers. They grow in high-elevation, wet, cold forests, where the brilliant flowers stand out against the generally dark green background. The plants form clusters of many elongate spindle-shaped pseudobulbs, which can each produce between one and three inflorescences.

The Orange Claw is different from its closest relatives, the spider orchids, which make up the majority of *Brassia* members, and is the only *Brassia* adapted to pollination by hummingbirds. If the flower parts are forced open, then the resulting shape is similar to that of the typical spider orchid, with elongate sepals and petals.

The flower of the Orange Claw is brilliant orange with lanceolate-acuminate sepals and petals, occasionally with purple-brown spotting at their base.

DISTRIBUTION

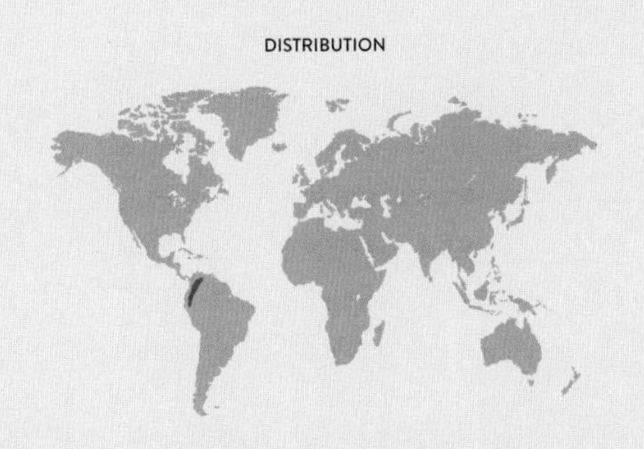

CUITLAUZINA PENDULA

AZTEC KING

LEXARZA, 1825

PLANT SIZE
8–12 × 5–10 in
(20–30 × 13–25 cm),
excluding pendent
inflorescences
10–30 in (25–76 cm) long

FLOWER SIZE
2 in (5 cm)

SUBFAMILY ~ Epidendroideae

TRIBE AND SUBTRIBE ~ Cymbidieae, Oncidiinae

NATIVE RANGE ~ West and central Mexico, in the states of Jalisco, Michoacan, and Sinaloa

HABITAT ~ Pine-oak forests, at 4,600–7,200 ft (1,400–2,200 m)

TYPE AND PLACEMENT ~ Epiphytic

CONSERVATION STATUS ~ Endangered as a result of overcollecting and habitat loss

FLOWERING TIME ~ May to October (late spring to fall)

The Aztec King has tightly clustered ovoid, laterally flattened pseudobulbs, each topped with two broad, leathery leaves. In the axils of a new growth, before the pseudobulb itself emerges, a pendent raceme carries 6–20, waxy, long-lived, lemon-scented flowers. In its natural habitat, the species can form large clumps that make an impressively beautiful display of pendent inflorescences.

The handsome genus is named for the Aztec king Cuitlahuatzin, or Cuitlahuac, (1476–1520), brother of Montezuma and a noted designer of early public gardens in Mexico; the common name similarly references the king. Although nothing has been published on the plant's pollination, the floral morphology—despite the white to pink flowers—is the same as many yellow-flowered *Oncidium* species common in Mexico. These are pollinated by bees seeking floral oil to mix with pollen as food for their young, which may also occur in the Aztec King.

The flower of the Aztec King has white, broad petals and sepals that are sometimes pink or pink-suffused. The petals are shortly clawed, and the usually pink lip is narrow at its yellow and red-spotted base with two broad, apical lobes. The column has wings and a hood.

DISTRIBUTION

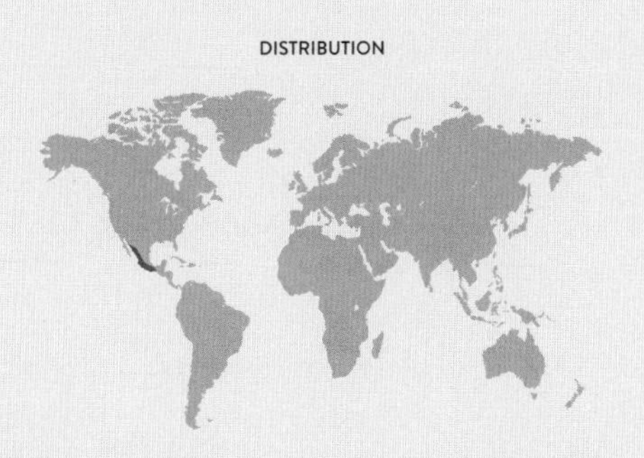

CYRTOCHILUM MACRANTHUM

GOLDEN CLOUD ORCHID

(LINDLEY) KRAENZLIN, 1917

PLANT SIZE
12–20 × 10–20 in (30–51 × 25–51 cm), excluding lateral, usually vinelike inflorescence, which can be 30–140 in (76–356 cm) long

FLOWER SIZE
4 in (10 cm)

SUBFAMILY ~ Epidendroideae

TRIBE AND SUBTRIBE ~ Cymbidieae, Oncidiinae

NATIVE RANGE ~ Northwestern South America (Colombia, Ecuador, northern Peru)

HABITAT ~ Cloud forests, up to 9,850 ft (3,000 m)

TYPE AND PLACEMENT ~ Epiphytic

CONSERVATION STATUS ~ Not assessed

FLOWERING TIME ~ Throughout the year, but mainly from fall to spring

The spectacular Golden Cloud Orchid produces clumps of conical pseudobulbs that are enveloped in several leaf-bearing sheaths and topped with two linear-oblong leaves. A lateral, stout, branched inflorescence rambles through the surrounding vegetation, each side branch carrying up to five large flowers. Although the genus name comes from the Greek words for "curved lip," *cyrtos* and *cheilos*, this characteristic is not true of this particular species.

The orchid is pollinated by *Centris* bees, which normally collect floral oils that are mixed with pollen from other plant species (not orchids) and fed to their young. However, the Golden Cloud Orchid does not produce floral oils, so this is a case of deceit. The 140 or so species of *Cyrtochilum* are distinguished from *Oncidium*, where they used to be placed, by their often elongate rhizomes, pseudobulbs that are round in cross section rather than flattened, and their rambling or twining inflorescences.

The flower of the Golden Cloud Orchid has large, yellowish-brown or yellow, spreading, clawed sepals and petals. The yellow lip is triangular, with its lateral tips blood-red to purple, and bears an intricate series of horns and protuberances. The column has two wings.

DISTRIBUTION

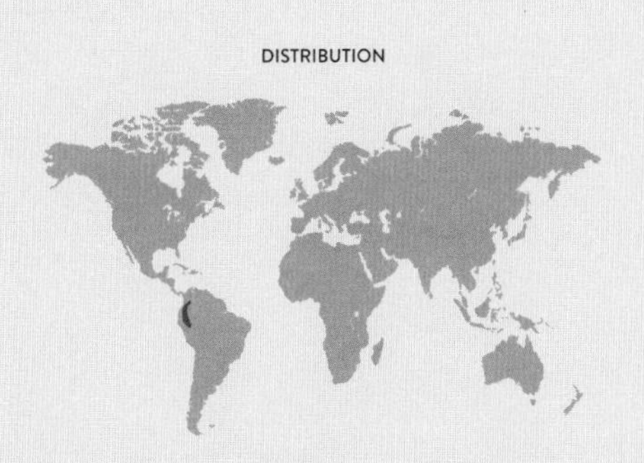

GOMESA FORBESII

SHINY FOREST SPRITE

(HOOKER) M. W. CHASE & N. H. WILLIAMS, 2009

PLANT SIZE
8–12 × 6–10 in (20–30 × 15–25 cm), excluding erect to arching inflorescence 10–16 in (25–41 cm) long

FLOWER SIZE
2 in (5 cm)

SUBFAMILY ~ Epidendroideae

TRIBE AND SUBTRIBE ~ Cymbidieae, Oncidiinae

NATIVE RANGE ~ Southeastern Brazil (Atlantic Forest)

HABITAT ~ Moderately wet forests, at 330–3,950 ft (100–1,200 m)

TYPE AND PLACEMENT ~ Epiphytic

CONSERVATION STATUS ~ Not threatened

FLOWERING TIME ~ September to October (spring)

The Shiny Forest Sprite produces flattened ovoid pseudobulbs with one or two narrowly lanceolate leaves on top. The inflorescence can bear up to 20 showy flowers. Originally a member of the large genus *Oncidium*, this and its related species were recently moved to *Gomesa*. The species name is in honor of the English botanist and plant collector John Forbes (1798–1823), while the common name refers to the glossy texture of the flowers, which are easily spotted in the wet Atlantic Forest of Brazil.

The species is pollinated by oil-collecting bees that mistake the flowers for those of the unrelated family Malpighiaceae. The lip callus appears to produce some oil, but it is not enough to reward the bees, which would normally collect the oil from Malpighiaceae blooms and mix it with pollen to feed their young.

The flower of the Shiny Forest Sprite has ovate, bright yellow sepals and petals that are heavily spotted with chestnut brown. The petals are much broader than the sepals. The lip has two small sidelobes and a widely spreading midlobe, again yellow with abundant chestnut-brown markings.

DISTRIBUTION

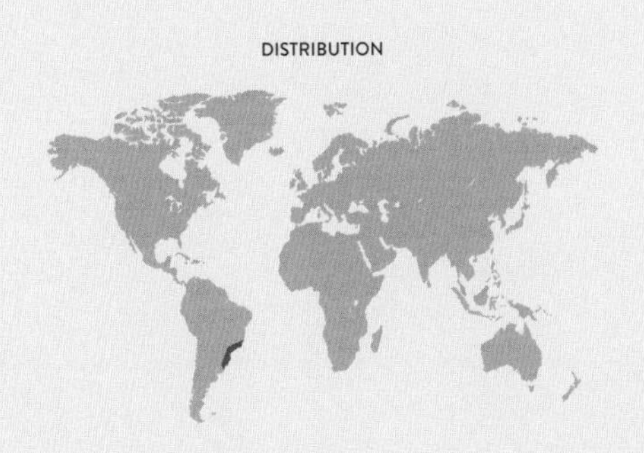

MILTONIOPSIS PHALAENOPSIS

BUTTERFLY-LIPPED ORCHID

(LINDEN & REICHENBACH FILS) GARAY & DUNSTERVILLE, 1976

PLANT SIZE
8–14 × 4–7 in (20–36 × 10–18 cm), including erect to arching inflorescence 7–12 in (18–30 cm) long, which is about as long as the leaves

FLOWER SIZE
2½ in (6.5 cm)

SUBFAMILY ~ Epidendroideae

TRIBE AND SUBTRIBE ~ Cymbidieae, Oncidiinae

NATIVE RANGE ~ Cundinamarca and Norte de Santander departments of Colombia

HABITAT ~ Wet forests, at 3,950–5,250 ft (1,200–1,600 m)

TYPE AND PLACEMENT ~ Epiphytic

CONSERVATION STATUS ~ Not threatened

FLOWERING TIME ~ At any time, but more often in late summer to fall

The Butterfly-lipped Orchid, which carries between three and five flowers per stem, has one of the most curiously patterned lips in the orchid family. Bees have been reported to pollinate other species in this genus, but the exact reasons for such a striking lip pattern, presumably an elaborate set of nectar guides, are unclear. Also, the lip is totally flat when most bee-pollinated species have a lip that to a degree surrounds the column. Pollination by bees at night has also been reported.

The genus name refers to the genus *Miltonia* (-*opsis* means "similar to") in which this species was previously classified. The species name —as in the well-known but unrelated genus of the same name—refers to the plant's labellum's resemblance to a moth or butterfly.

The flower of the Butterfly-lipped Orchid is almost perfectly flat and has white (rarely pale pink) sepals and petals. The extraordinary lip is large and shaped like a butterfly with a complex set of purple-red and yellow markings on a white background.

DISTRIBUTION

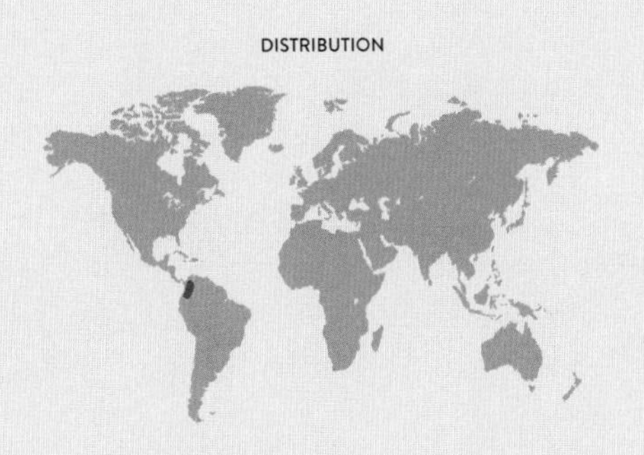

ONCIDIUM HARRYANUM

HARRY'S MOUNTAIN ORCHID

(REICHENBACH FILS) M. W. CHASE & N. H. WILLIAMS, 2008

PLANT SIZE
10–18 × 8–12 in (25–46 × 20–30 cm), excluding arching inflorescence 12–18 in (30–46 cm) tall

FLOWER SIZE
4 in (10 cm)

SUBFAMILY ~ Epidendroideae

TRIBE AND SUBTRIBE ~ Cymbidieae, Oncidiinae

NATIVE RANGE ~ Western South America, from Antioquia department in Colombia to Peru

HABITAT ~ Edges of forests, at 2,300–9,850 ft (700–3,000 m)

TYPE AND PLACEMENT ~ Epiphytic

CONSERVATION STATUS ~ Not formally assessed

FLOWERING TIME ~ December to March (summer to fall)

Harry's Mountain Orchid produces tight clusters of ovoid-elliptic, ribbed, and laterally flattened pseudobulbs, which are subtended by several leaf-bearing sheaths and topped by two oblong-elliptic to narrowly oblong leaves. From the base of a pseudobulb, an almost upright stem carries up to a dozen fragrant, long-lasting, waxy flowers. The species and common name refer to Harry James Veitch (1840–1924), a prominent English horticulturalist and one of the early promoters of the Great International Horticultural Exhibition, which subsequently became the Chelsea Flower Show.

Oncidium harryanum most likely attracts bees for pollination, but this has not been observed in the wild. There is no obvious reward for a pollinator, so the process must involve deceit for attraction of the insects. The species is also among those formerly placed in the genus *Odontoglossum*.

The flower of Harry's Mountain Orchid has yellowish-green to tan sepals and petals that have a purple to reddish overlay. The lip has short sidelobes, several callus projections and hairs, and is broadly lobed with purple and white markings.

DISTRIBUTION

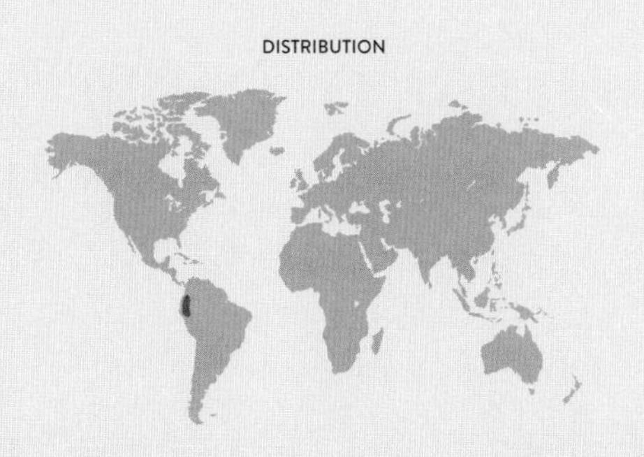

ONCIDIUM NOBILE

NOBLE SNOW ORCHID

(REICHENBACH FILS) M. W. CHASE & N. H. WILLIAMS, 2008

PLANT SIZE
10–18 × 12–18 in (25–46 × 30–46 cm), excluding arching lateral inflorescence 14–24 in (36–61 cm) long

FLOWER SIZE
2½ in (6.5 cm)

SUBFAMILY ~ Epidendroideae

TRIBE AND SUBTRIBE ~ Cymbidieae, Oncidiinae

NATIVE RANGE ~ Colombia

HABITAT ~ Cloud forests, at 6,600–7,900 ft (2,000–2,400 m)

TYPE AND PLACEMENT ~ Epiphytics

CONSERVATION STATUS ~ Not assessed

FLOWERING TIME ~ May to August (spring to summer)

In its cool foggy forest habitats the Noble Snow Orchid remains well hidden until it opens its large, white, spectacular flowers, which have long been held in high regard by orchid fanciers. In the past, the species was placed in the genus *Odontoglossum*, which is based on the Greek words *odon*, "tooth," and *glossa*, "tongue," alluding to the toothlike lip projections.

The plant produces a faint fragrance that at times smells sweet and at other times somewhat sharp and disagreeable. The partial fusion of the lip to the column creates a false nectary, which means that to reach it a bee (most likely a bumblebee) must overcome the impediments created by the complex toothlike lip callus. In struggling to get past, the bee contacts the sticky disc that projects over the lip callus and pulls out the pollinia.

The flower of the Noble Snow Orchid has broadly lanceolate, projecting, white sepals and petals. The lip is also white and weakly trilobed, the broad apical lobe itself bilobed with a complex, red-striped, yellow callus. The column is winged and red spotted.

DISTRIBUTION

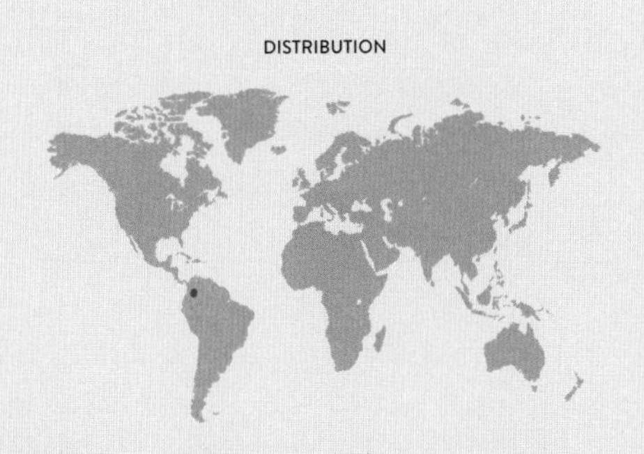

PSYCHOPSIELLA LIMMINGHEI

LITTLE BUTTERFLY ORCHID

(E. MORREN EX LINDLEY) LÜCKEL & BRAEM, 1982

PLANT SIZE
1–2 × 2–3 in
(2.5–5 × 5–8 cm),
excluding short inflorescence
2–3 in (5–8 cm) tall

FLOWER SIZE
1¼–1½ in (3–4 cm)

SUBFAMILY ~ Epidendroideae

TRIBE AND SUBTRIBE ~ Cymbidieae, Oncidiinae

NATIVE RANGE ~ Brazil (near Rio de Janeiro)

HABITAT ~ Seasonally arid areas, often on cactus

TYPE AND PLACEMENT ~ Epiphytic

CONSERVATION STATUS ~ Not assessed

FLOWERING TIME ~ Spring to fall

The miniature creeping Little Butterfly Orchid, the only species in its genus, is, as the name suggests, a close relative of the plants of the genus Psychopsis, though highly reduced in size. Small, flattened, overlapping heart-shaped pseudobulbs with red-veined and mottled leaves clasp their substrate.

The intricate flowers are large, considering the tiny plants from which they emerge. They are somewhat different from typical Psychopsis plants in not bearing petals and sepals resembling antennae and appearing more like the oil-rewarding mimics of the Oncidiinae subtribe. The plants also differ from Psychopsis species in that the inflorescences do not produce flowers successively. Indeed, all Psychopsis species were once treated as a section (Glanduligera) of the genus Oncidium, until taxonomists realized their many differences.

The flower of the Little Butterfly Orchid is pale yellow with reddish-brown patterning on all segments and lip, particularly near the lip margin, where a zone of spots occurs. The column bears feathery arms on each side that have a minute gland at each tip.

DISTRIBUTION

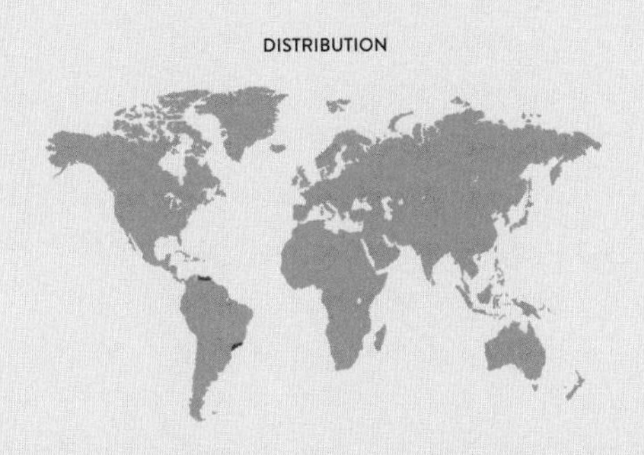

ROSSIOGLOSSUM GRANDE

TIGER'S MOUTH

(LINDLEY) GARAY & G. C. KENNEDY, 1976

PLANT SIZE
8–12 × 8–10 in
(20–30 × 20–25 cm),
excluding erect-arching
inflorescence, which is
longer than the leaves,
10–16 in (25–41 cm) long

FLOWER SIZE
7–9 in (18–23 cm)

SUBFAMILY ~ Epidendroideae

TRIBE AND SUBTRIBE ~ Cymbidieae, Oncidiinae

NATIVE RANGE ~ Mexico (Chiapas state), Guatemala, and Belize

HABITAT ~ Deciduous, cool wet forests, at 4,920–8,900 ft (1,500–2,700 m)

TYPE AND PLACEMENT ~ Epiphytic

CONSERVATION STATUS ~ Not assessed

FLOWERING TIME ~ November to January

The striking, large-flowered Tiger's Mouth, named for its coloration and the pair of prominent teeth on the callus in the center of the lip (*boca del tigre* in Spanish), is possibly the best-known member of the genus *Rossioglossum*. Most species occur at higher elevations in seasonally dry deciduous forests, where their large size and bright colors with a glossy, lacquered finish make an impressive display.

Like many other species in the Oncidiinae subtribe, these orchids engage in oil-reward deception, mimicking other, more common flowers that actually reward *Centris* bees with oil for their pollination services. There seems to be a small amount of oil produced on the complex shiny lip callus, but it would not appear that there is enough oil production to actually serve as reward for the bees.

The flower of the Tiger's Mouth has bright yellow sepals and petals overlaid with reddish-brown barring that is almost solid at the base. The lip is paler yellow, with a prominent reddish-brown-barred lip callus with several teeth. The column is yellow and winged.

DISTRIBUTION

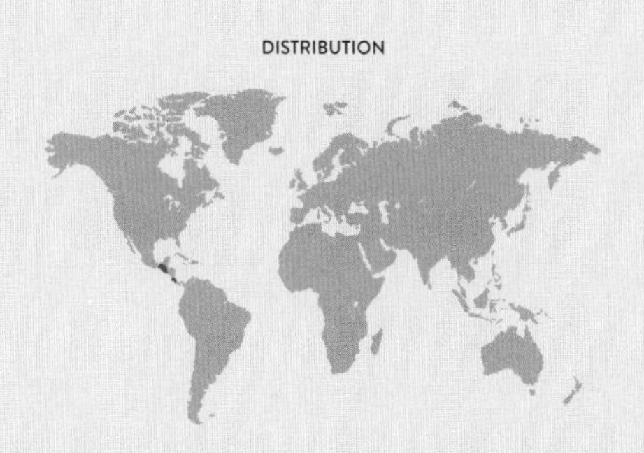

TRICHOPILIA SUAVIS

PINK-SPOTTED HOOD ORCHID

LINDLEY & PAXTON, 1850

PLANT SIZE
10–18 × 4–6 in (25–46 × 10–15 cm), excluding the pendent inflorescence, which is 4–8 in (10–20 cm) long

FLOWER SIZE
4–4¾ in (10–12 cm)

SUBFAMILY ~ Epidendroideae

TRIBE AND SUBTRIBE ~ Cymbidieae, Oncidiinae

NATIVE RANGE ~ Costa Rica to northern Colombia

HABITAT ~ Montane, seasonally dry forest margins

TYPE AND PLACEMENT ~ Epiphytic

CONSERVATION STATUS ~ Vulnerable to poaching

FLOWERING TIME ~ February to April

Despite bearing large, intensely fragrant, and vividly colored blossoms, the Pink-spotted Hood Orchid is, surprisingly, only occasionally cultivated. The medium-sized plants grow low on thick forest branches, where the pendent spikes attract male euglossine bees, although it is not clear if they are collecting floral fragrance compounds, as is typical for these insects.

The genus is closely related to *Psychopsis* and *Psychopsiella*, a group of species with oil-producing bee-attracting blooms that could not differ more from those of *Trichopilia*—evidence that pollination drives evolution of floral features, even in closely related orchids. The common name of the genus refers to the hood on the apex of the column that covers the anther cap, a reference that occurs in the genus name also, from the Greek word *pilos*, "felt," a material used to make hoods.

The flower of the Pink-spotted Hood Orchid has white or cream tepals (often marked with pale pink) and a tubular lip, typical of many orchids pollinated by euglossine bees. The labellum is crisped around its margin and has vivid pink to red-pink markings on its interior surface.

DISTRIBUTION

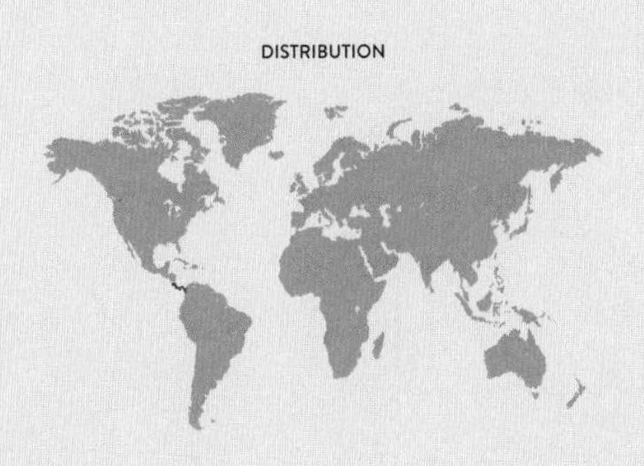

PERISTERIA ELATA

HOLY GHOST ORCHID

HOOKER, 1831

PLANT SIZE
25–40 × 20–30 in (64–102 × 51–76 cm), excluding erect, basal inflorescence, which is much taller than the leaves, 35–55 in (89–140 cm) long

FLOWER SIZE
2 in (5 cm)

SUBFAMILY ~ Epidendroideae

TRIBE AND SUBTRIBE ~ Cymbidieae, Coeliopsidinae

NATIVE RANGE ~ Central America and northwestern South America, Costa Rica to Ecuador

HABITAT ~ Humid, deciduous mountain forests, at about 3,300–3,600 ft (100–1,100 m)

TYPE AND PLACEMENT ~ Lithophytic or terrestrial, or epiphytic at the base of mossy tree trunks

CONSERVATION STATUS ~ Not formally assessed and still locally abundant, but threatened by overcollecting

FLOWERING TIME ~ July to August (summer)

The large Holy Ghost Orchid has conical or ovoid pseudobulbs enveloped in papery sheaths and topped with up to four broadly lanceolate, folded leaves. From the base of the pseudobulb, the plant produces a rigidly erect inflorescence, with 10–15 waxy, campanulate flowers that have a strong fragrance, typical of male euglossine bee pollination.

The lip is hinged, and the weight of the bee, *Euplusia concava*, when landing on the lip causes the hinge to tip and the bee to be thrown against the column, where it picks up the pollinarium while struggling to get free. The orchid, which is the national flower of Panama, has a shape said to resemble a white dove sitting on a nest, hence its other common name, the Dove Orchid. The generic name is close to the Greek word for "dove"—*peristéri*.

The flower of the Holy Ghost Orchid has broad, white sepals and petals that form an enclosed cup around the saccate, white lip and beaked column. The lip has two lateral lobes that are spotted purple.

DISTRIBUTION

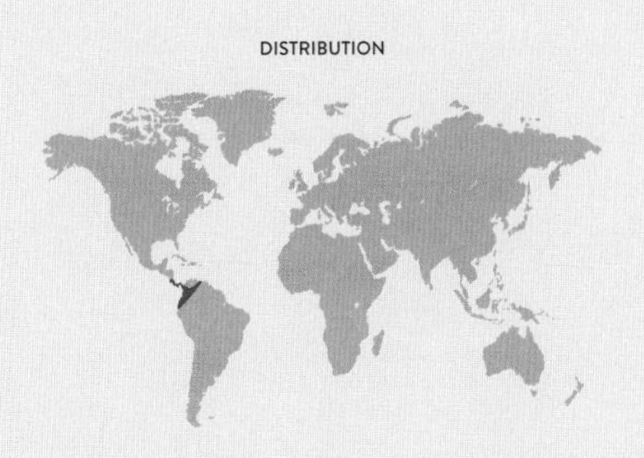

CORYANTHES SPECIOSA

BEAUTIFUL BUCKET ORCHID

HOOKER, 1831

PLANT SIZE
10–18 × 8–12 in (25–46 × 20–31 cm), excluding pendent inflorescence 8–14 in (20–36 cm) long

FLOWER SIZE
5 in (13 cm)

SUBFAMILY ~ Epidendroideae

TRIBE AND SUBTRIBE ~ Cymbidieae, Stanhopeinae

NATIVE RANGE ~ Trinidad and Tobago, French Guiana, Surinam, Guyana, Venezuela, Peru, and Brazil

HABITAT ~ Low to mid elevation tropical wet forests and cultivated guava and citrus groves

TYPE AND PLACEMENT ~ Epiphytic

CONSERVATION STATUS ~ Not threatened

FLOWERING TIME ~ Summer

The Beautiful Bucket Orchid displays one of the most remarkable and mind-boggling pollination mechanisms. Attracted to the large, intricate, pendent blooms by an indescribably complex aroma, fragrance-collecting male euglossine bees accidentally fall into the bucket-shaped lip, which is filled with liquid from faucet-like glands at the base of the column. The only way out of the bucket is via a small opening near the rear of the flower, where the bees pick up (or deposit) pollinia as they squeeze their way through it.

The genus name is derived from the Greek words *korys*, meaning "helmet," and *anthos*, "flower," a reference to the shape of the lip. The species name, *speciosa*, is Latin for "showy." The plants grow in aerial ant colonies, where formic acid produced by the hosts creates acidic conditions, to which the plants are adapted.

The flower of the Beautiful Bucket Orchid is large, complex, and variable in color but generally with various tawny shades of yellow to brown, often spotted with reddish-brown. The sepals and petals are winglike and reflexed, leaving the pouch-like lip as the prominent floral feature.

DISTRIBUTION

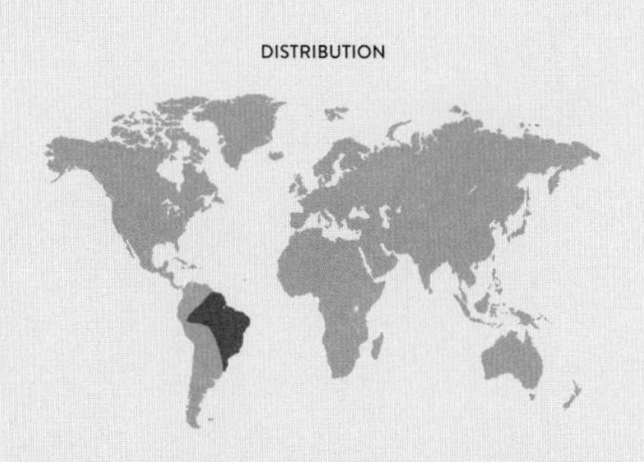

CYCNOCHES CHLOROCHILON

GREEN-LIPPED SWAN ORCHID

KLOTZSCH, 1838

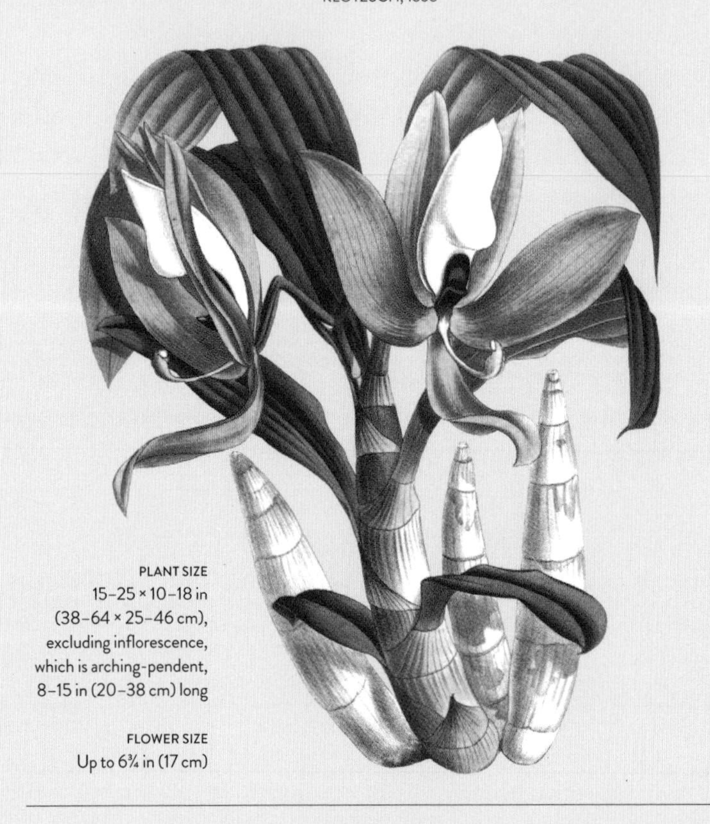

PLANT SIZE
15–25 × 10–18 in (38–64 × 25–46 cm), excluding inflorescence, which is arching-pendent, 8–15 in (20–38 cm) long

FLOWER SIZE
Up to 6¾ in (17 cm)

SUBFAMILY ~ Epidendroideae

TRIBE AND SUBTRIBE ~ Cymbidieae, Catasetinae

NATIVE RANGE ~ Southeastern Panama, Colombia, and Venezuela

HABITAT ~ Lowland wet tropical forests, at around 1,300–2,600 ft (400–800 m)

TYPE AND PLACEMENT ~ Epiphytic

CONSERVATION STATUS ~ Not threatened

FLOWERING TIME ~ Fall

The Green-lipped Swan Orchid, so-called for the long, graceful curve of its column, is one of the larger and more robust species in a genus of mostly warm-growing, low-elevation deciduous epiphytes with showy, fragrant, and sexually dimorphic flowers. The genus name, *Cycnoches*, is from the Greek words *kyknos*, "swan," and *auchen*, "neck."

Elongating rapidly and gradually fattening their leafy cigar-shaped pseudobulbs during the rainy season, plants become completely leafless during the dry season. Flower spikes usually appear near the apex of mature pseudobulbs just before the leaves are shed. With similar-sized male and female flowers, this species may be the least sexually dimorphic of the genus *Cycnoches*. Other species in related genera, such as *Catasetum*, can bear numerous large and colorful male blooms with projectile pollinia. These dimorphic genera are all pollinated by fragrance-collecting male euglossine bees.

The flower of the Green-lipped Swan Orchid is non-resupinate, waxy in appearance, and star-shaped. It is usually pale green to greenish-orange with a large white lip and darker green lip callus.

DISTRIBUTION

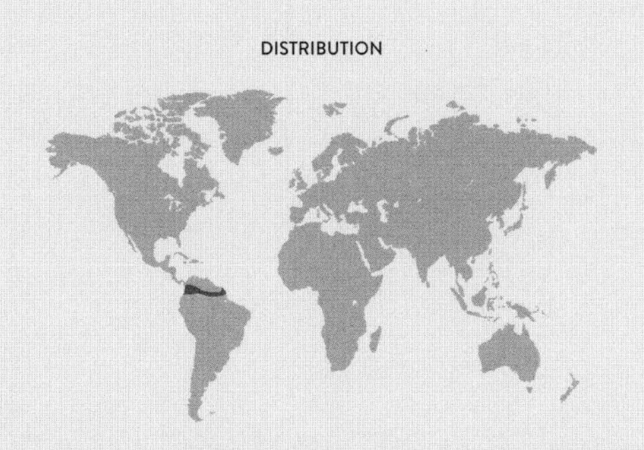

LUEDDEMANNIA PESCATOREI

GOLDEN-BROWN FOXTAIL ORCHID

(LINDLEY) LINDEN & REICHENBACH FILS, 1854

PLANT SIZE
18–36 × 12–25
(46–91 × 30–64 cm), excluding
pendent inflorescence
20–40 in (51–102 cm) long

FLOWER SIZE
1 in (2.5 cm)

SUBFAMILY ~ Epidendroideae

TRIBE AND SUBTRIBE ~ Cymbidieae, Stanhopeinae

NATIVE RANGE ~ Northwestern South America, south to Peru

HABITAT ~ Wet forests at 3,300–3,950 ft (1,000–1,200 m)

TYPE AND PLACEMENT ~ Epiphytic, rarely terrestrial on steep slopes

CONSERVATION STATUS ~ Least concern

FLOWERING TIME ~ July to August (summer)

The large Golden-brown Foxtail Orchid produces a cluster of egg-shaped, grooved pseudobulbs that carry two to four elliptic, plicate, clasping leaves. It has a long, pendulous inflorescence that grows directly underneath the plant and is densely set with numerous (up to 50) sweetly scented flowers, each subtended by an elliptic bract. The bracts, flower stalks, and outer surfaces of the sepals are covered with black-brown globular bodies, for which the function is unknown.

The species is widespread from Venezuela to Peru, but collections are scarce. It can be found at lower elevations around Machu Picchu, where numerous tourists will have encountered it. The pollinator of the orchid is not known, but the shape and fragrance indicate some type of bee as a visitor.

The flower of the Golden-brown Foxtail Orchid has brownish-orange sepals and similar but smaller, forward-curved, bright yellow petals, together forming a tube with the bright yellow column and lip. The lip has an arrowhead-shaped midlobe.

DISTRIBUTION

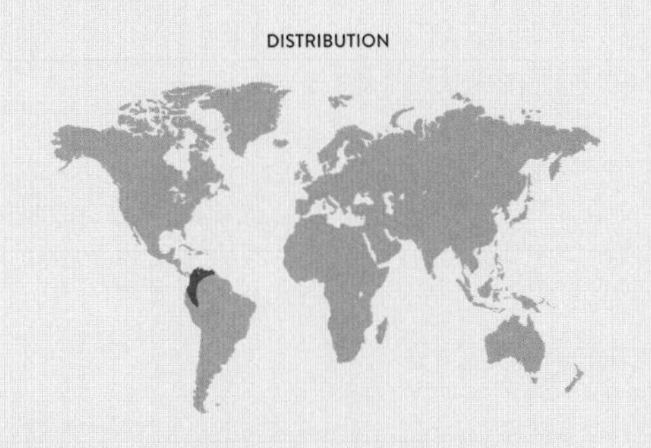

PAPHINIA CRISTATA

WHITE-BEARDED STAR ORCHID

(LINDLEY) LINDLEY, 1843

PLANT SIZE
8–12 × 6–10 in (20–30 × 15–25 cm), excluding pendent inflorescence 3–8 in (8–20 cm) long

FLOWER SIZE
4 in (10 cm)

SUBFAMILY ~ Epidendroideae

TRIBE AND SUBTRIBE ~ Cymbidieae, Stanhopeinae

NATIVE RANGE ~ Northern South America

HABITAT ~ Wet forests in shade, mostly on moss-covered understory trees, from sea level to 4,920 ft (1,500 m)

TYPE AND PLACEMENT ~ Epiphytic

CONSERVATION STATUS ~ Not assessed

FLOWERING TIME ~ April to May

Relatively small plants with comparatively large and stunning flowers, *Paphinia* species thrive in sultry, humid conditions. Their star-shaped flowers appear on sharply pendent stems and can occur in great profusion. Most of the 15 described species have only been discovered since the 1980s.

Although closely related to the powerfully fragrant species of *Stanhopea*, which achieve pollination by attracting male euglossine bees that harvest fragrance compounds, *Paphinia* species are not particularly fragrant. Nonetheless, they do manage to attract euglossine bees, presumably because there are compounds present that human noses (and gas chromatographs) cannot detect. The lip of the White-bearded Star Orchid is adorned with a filamentous crest or comb from which it gets its species and common names, from the Latin word *cristatus*, meaning "crested." *Paphia* was a local Cypriot name for Aphrodite.

The flower of the White-bearded Star Orchid has white parts so extensively overlaid with maroon markings that they appear almost solidly colored. The lip has an impressive apical, white beard, and the column is yellow with an inflated apex.

DISTRIBUTION

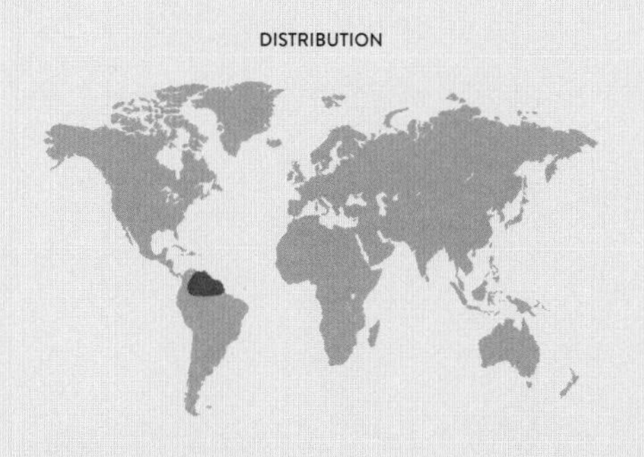

HUNTLEYA MELEAGRIS

GUINEA FOWL ORCHID

LINDLEY, 1837

PLANT SIZE
12–20 × 12–25 in
(30–51 × 30–64 cm),
including short, single-flowered inflorescences
5–8 in (13–20 cm) long

FLOWER SIZE
4¾–6 in (12–15 cm)

SUBFAMILY ~ Epidendroideae

TRIBE AND SUBTRIBE ~ Cymbidieae, Zygopetalinae

NATIVE RANGE ~ Venezuela, Guyana, Brazil, Peru, Bolivia, and Trinidad and Tobago

HABITAT ~ Wet forests, at 2,000–4,300 ft (600–1,300 m)

TYPE AND PLACEMENT ~ Epiphytic

CONSERVATION STATUS ~ Not threatened

FLOWERING TIME ~ Spring to summer

The stately, often stunning Guinea Fowl Orchid has large, showy, and aromatic flowers with distinctive characteristics. Leaves are arranged in a graceful fan shape and make a perfect frame for the spectacular star-shaped flowers that emerge singly on short stems from the leaf axils.

The flowers are large and flat as well as colorful, usually with concentric spotted patterns and a glossy sheen, like the feathers of a guinea fowl, hence the common name. Male euglossine bees foraging for fragrance compounds pollinate the species. Its labellum bears a crest of rigid papillae that subtend the column and probably steer the pollinator to its desired position on the flower, where pollinia will adhere effectively. In nature, the lips of the flowers can be heavily damaged by the foraging bees in their efforts to collect the fragrances.

The flower of the Guinea Fowl Orchid is star-shaped, with broad petals and sepals that taper to an acuminate tip. It is generally mottled purplish-brown but set off by a white and yellow center and a whitish, brown-tipped lip, with a crest of curving papillae.

DISTRIBUTION

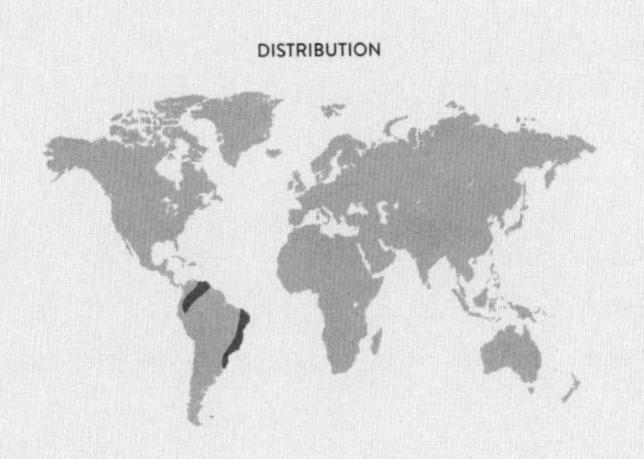

PABSTIA JUGOSA

HAWK ORCHID

(LINDLEY) GARAY, 1873

PLANT SIZE
9–15 × 6–12 in (23–38 × 15–30 cm), excluding inflorescence, which is nearly always shorter than the leaves

FLOWER SIZE
2 in (5 cm)

SUBFAMILY ~ Epidendroideae

TRIBE AND SUBTRIBE ~ Cymbidieae, Zygopetalinae

NATIVE RANGE ~ Eastern Brazil, Serra do Mar from Rio de Janeiro to Espírito Santo

HABITAT ~ Atlantic rain forests in shady and humid places, at up to 2,300 ft (700 m)

TYPE AND PLACEMENT ~ Epiphytic or lithophytic

CONSERVATION STATUS ~ Not formally assessed, but probably threatened by deforestation

FLOWERING TIME ~ October to January (spring to early summer)

The Hawk Orchid produces a cluster of ovoid, slightly compressed, nearly smooth pseudobulbs, each bearing two or three elongate-lanceolate leaves. On new growths, an upright inflorescence grows, with up to four highly fragrant flowers. The genus is closely related to *Zygopetalum*, from which it differs in minor characteristics.

The purple-spotted petals contrast strongly with the white sepals, and together with the column, which forms the "head," these resemble a bird of prey in flight, hence the common name. The genus name honors the renowned Brazilian orchidologist, Guido Pabst (1914–80), founder of the Herbarium Bradeanum in Rio de Janeiro. Pollination of this species has not been studied, but it is likely that the pollinators are male euglossine bees seeking floral fragrance compounds.

The flower of the Hawk Orchid has fleshy, spreading, white sepals and petals, the latter with reddish-purple spots and bars. The white, blue-marked lip is fan-shaped with the edges curved down. The anther cap is large and conspicuous.

DISTRIBUTION

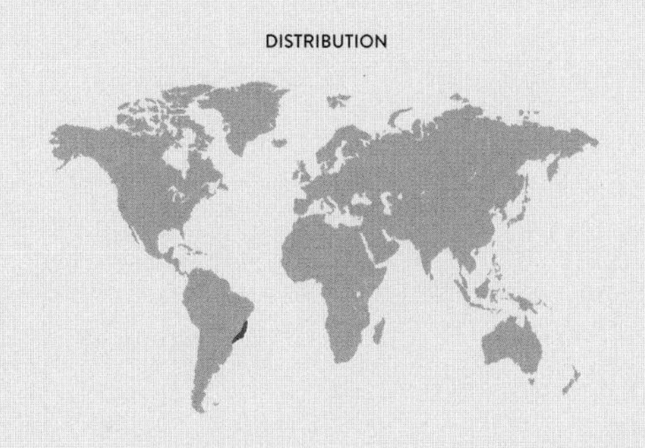

ZYGOPETALUM MACULATUM

SPOTTED CAT OF THE MOUNTAIN

(KUNTH) GARAY, 1970

PLANT SIZE
15–25 × 12–18 in (38–64 × 30–46 cm), excluding erect-arching inflorescence 20–35 in (51–89 cm) tall

FLOWER SIZE
1¾ in (4.5 cm)

SUBFAMILY ~ Epidendroideae

TRIBE AND SUBTRIBE ~ Cymbidieae, Zygopetalinae

NATIVE RANGE ~ Northern Peru to Bolivia and eastern Brazil

HABITAT ~ Mountain slopes and ridges with shrubs and rocks, sometimes on roadside banks, at 3,300–8,200 ft (1,000–2,500 m)

TYPE AND PLACEMENT ~ Terrestrial in wet, mossy sites among rocks

CONSERVATION STATUS ~ Widespread and locally abundant, but not assessed

FLOWERING TIME ~ September to October (spring), but potentially throughout the year

The Spotted Cat of the Mountain can grow into large clumps of tightly clustered, ovoid pseudobulbs with some basal leaf-bearing bracts and two to three apical linear leaves. These plants, which generally grow in places with some shade from nearby trees, can be spectacular, with one or two tall inflorescences per pseudobulb carrying up to 25 flowers. The genus name is based on the Greek words *zygon*, "yoke," and *petalon*, "petal," referring to the way the lip appears to tie together, or yoke, the petals. The species name, which means "spotted" in Latin, refers, as does the common name, to the spots on the sepals and petals.

The species has a pleasingly sweet fragrance dominated by the compounds that are typically associated with fragrance-collecting male euglossine bees. Pollination, though, has yet to be studied in the wild.

The flower of the Spotted Cat of the Mountain has green sepals and petals similar in size and shape, well covered with reddish spots and bars. The lip has a single large lobe, which is white with purple veins, and a large, raised basal callus.

DISTRIBUTION

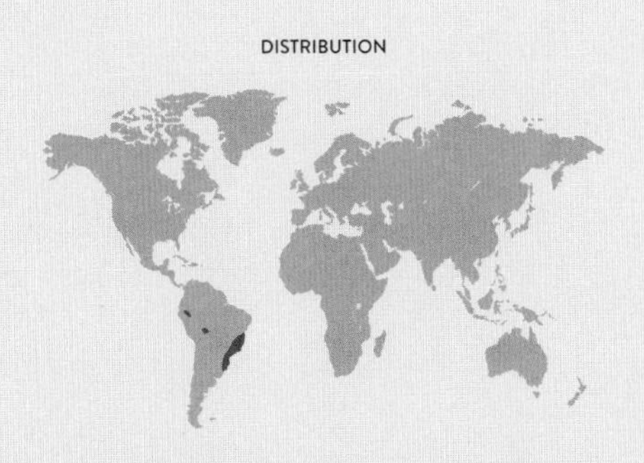

CALYPSO BULBOSA

FAIRY SLIPPER

(LINNAEUS) OAKES, 1842

PLANT SIZE
Usually only one basal leaf 2–3 in (5–8 cm) long, held flat on the soil surface, with an erect inflorescence 4–6 in (10–15 cm) tall

FLOWER SIZE
1⅜–1½ in (3.5–4 cm)

SUBFAMILY ~ Epidendroideae

TRIBE AND SUBTRIBE ~ Epidendreae, Calypsoinae

NATIVE RANGE ~ Circumboreal—northern Europe, Asia, and North America

HABITAT ~ Boreal, mostly evergreen forests

TYPE AND PLACEMENT ~ Terrestrial

CONSERVATION STATUS ~ Not threatened

FLOWERING TIME ~ Early spring, as snow melts, later farther north

A beautiful spring-flowering species from the far northern latitudes of our globe, *Calypso bulbosa* was named in honor of the seductive sea nymph, Calypso, from Homer's *Odyssey*. It has several distinctive regional varieties over its vast range. Preferring shady habitats in old-growth evergreen forests, the diminutive plants bear a single, dark green, ovoid, pleated leaf with a short inflorescence and one vanilla-scented bloom. When Linnaeus originally described the species, it was treated in the genus *Cypripedium* due to its lip, which is superficially similar to those of the lady's slipper orchids.

As the species name implies, there is a subterranean corm, which was collected as food by Native Americans in northwestern North America. The flowers of some forms have yellow tufts at the opening of their lip, which are thought to be mimicking pollen. Bumblebees are their usual pollinators.

The flower of the Fairy Slipper varies by region, but sepals and petals are generally pink, sometimes white. In variety *americana* the pouch-like lip is streaked internally with vivid maroon and has yellow hairs. Other forms have heavily red-spotted lips with a white tuft of hairs.

DISTRIBUTION

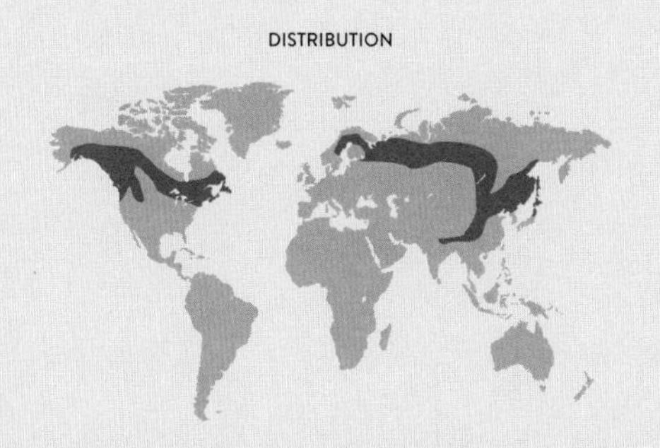

COELIA BELLA

BEAUTIFUL EGG ORCHID

(LEMAIRE) REICHENBACH FILS, 1861

PLANT SIZE
10–18 × 6–10 in (25–46 × 15–25 cm), with an erect-arching inflorescence shorter than the leaves, 6–10 in (15–25 cm) long

FLOWER SIZE
2 in (5 cm)

SUBFAMILY ~ Epidendroideae

TRIBE AND SUBTRIBE ~ Epidendreae, Calypsoinae

NATIVE RANGE ~ Mexico, Honduras, and Guatemala

HABITAT ~ Rain forests in partial shade

TYPE AND PLACEMENT ~ Mostly terrestrial, sometimes epiphytic and lithophytic

CONSERVATION STATUS ~ Not threatened

FLOWERING TIME ~ Summer

The tough, sturdy, and adaptable Beautiful Egg Orchid is common and abundant in many locations. Capable of growing high up in trees as an epiphyte, it is even more plentiful as a terrestrial or lithophyte over much of its range. It has shiny pseudobulbs that are ovoid, or egg-shaped—hence the common name—with between three and five lanceolate apical leaves. Inflorescences emerge from the base of the plants and bear colorful flowers that emit a sweet fragrance.

Coelia bella usually has 6–12 blooms per spike, and these are often nestled in among the pseudobulbs or hidden beneath plentiful foliage. The pleasing scent implies a bee pollination syndrome, which makes sense in terms of floral morphology, although no studies have so far documented what sort of bees (which receive no reward) might be involved.

The flower of the Beautiful Egg Orchid is crystalline in texture, usually white, with sepals larger than the petals and tipped with brilliant rose. The lip, usually sulfur yellow, is gullet-shaped with a recurving midlobe.

DISTRIBUTION

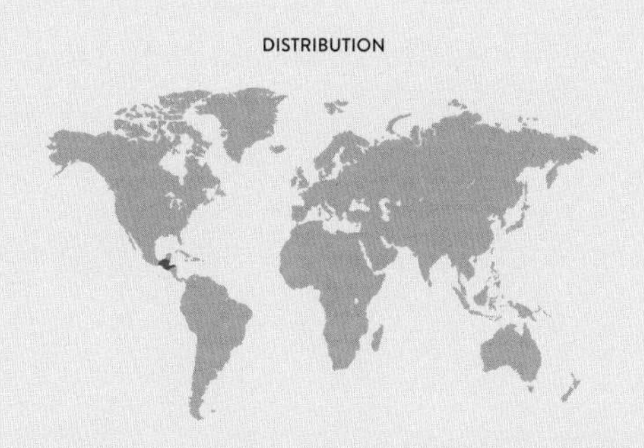

ARPOPHYLLUM GIGANTEUM

CANDLESTICK ORCHID

HARTWEG EX LINDLEY, 1840

PLANT SIZE
20–30 × 4–5 in
(51–76 × 10–13 cm),
excluding terminal erect
inflorescence, which extends
beyond the leaves, 6–12 in
(15–30 cm) long

FLOWER SIZE
¼ in (0.8 cm)

SUBFAMILY ~ Epidendroideae

TRIBE AND SUBTRIBE ~ Epidendreae, Laeliinae

NATIVE RANGE ~ Mexico, Guatemala, Belize, El Salvador, Honduras, Nicaragua, Costa Rica, Colombia, Venezuela, and Jamaica

HABITAT ~ Seasonally dry foothill forests, at 2,600–4,900 ft (800–1,500 m)

TYPE AND PLACEMENT ~ Epiphytic

CONSERVATION STATUS ~ Not threatened

FLOWERING TIME ~ Late winter to early spring

Commonly encountered in the foothills of the mountain systems of Central America, the Candlestick Orchid is well known to locals, who call it *pico de curvo* (raven's beak) or *masorquilla* (little corn orchid) and often use its leaves for medicinal purposes, including the treatment of dysentery. The plant has long, narrow, scimitar-like leaves and beautiful, jewellike, spirally arranged flowers on long, showy inflorescences that arise from a sheath at the pseudobulb apex. The genus name, from the Greek words *arpe*, "sickle," and *phyllon*, "leaf," refers to the shape of the leaves.

Despite the orchid's familiarity and popularity, pollination has not been documented. However, due to the vibrant colors, copious nectar, and darkly colored pollen masses, the probable pollinators are birds, which are thought to be less likely to remove dark pollinia from their beaks.

The flower of the Candlestick Orchid is non-resupinate and forms part of a tight cylindrical spiral of blooms. The tepals are various shades of pink or pale purple with a darker, rounded labellum of often-brilliant magenta or purple and a dark purple column.

DISTRIBUTION

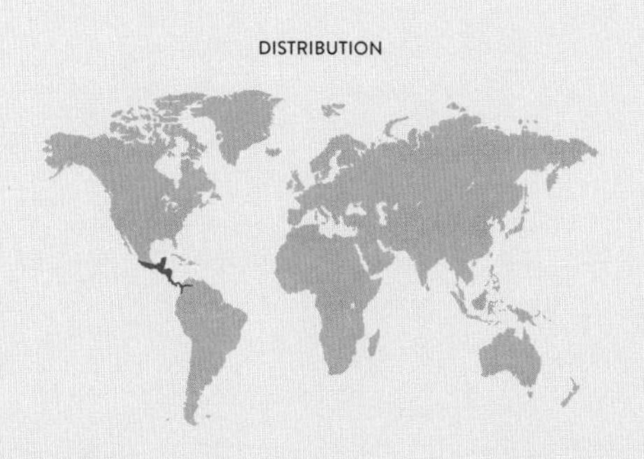

BRASSAVOLA NODOSA

LADY OF THE NIGHT

(LINNAEUS) LINDLEY, 1831

PLANT SIZE
5–9 × 1–1½ in (13–23 × 2.5–3.8 cm), excluding erect to arching 6–10 in (15–25 cm) long inflorescence

FLOWER SIZE
4 in (10 cm)

SUBFAMILY ~ Epidendroideae

TRIBE AND SUBTRIBE ~ Epidendreae, Laeliinae

NATIVE RANGE ~ Widespread in lowland forests, from Mexico to Brazil and the Caribbean

HABITAT ~ Seasonally dry forests, often near the coast, sea level to 1,640 ft (500 m)

TYPE AND PLACEMENT ~ Epiphytic

CONSERVATION STATUS ~ Not threatened

FLOWERING TIME ~ Throughout the year

The Lady of the Night produces clusters of thick, almost round pencil-like leaves, sometimes covering the trunks of trees with what looks like a massive shaggy carpet. Between four and twelve flowers are produced in tight clusters that emit a wonderful sweet nocturnal fragrance. Although its common name implies a link between its scent and the cheap perfume worn by women of ill repute, the fragrance is far more alluring, especially to the pollinating moths it is attempting to seduce. The genus name is after the Italian physician Antonio Musa Brasavola (1500–55).

Although not bearing a long spur like other moth-pollinated flowers such as species of the genus *Angraecum*, including the Madagascar Comet Orchid, the members of *Brassavola* have instead a nectar tube that runs underneath the ventral surface of the ovary. This serves the same purpose as a spur.

The flower of the Lady of the Night has green to whitish-cream, narrowly lanceolate, spidery sepals and petals. Its large, prominent tubular lip flares into a heart shape, sometimes with a few reddish spots deep in the throat.

DISTRIBUTION

CATTLEYA ACLANDIAE

LADY ACKLAND'S ORCHID

LINDLEY, 1840

PLANT SIZE
4–8 × 3–6 in (10–20 × 8–15 cm), excluding erect inflorescence 2–3 in (5–8 cm) long

FLOWER SIZE
2⅜–4 in (6–10 cm)

SUBFAMILY ~ Epidendroideae

TRIBE AND SUBTRIBE ~ Epidendreae, Laeliinae

NATIVE RANGE ~ State of Bahia (Brazil)

HABITAT ~ Hot lowland scrub near the beach

TYPE AND PLACEMENT ~ Epiphytic

CONSERVATION STATUS ~ Threatened by collecting and habitat degradation

FLOWERING TIME ~ Summer to fall

A dwarf member of its genus, Lady Ackland's Orchid clings to rough-barked trees at sea level. It bears two leaves at the apex of each slender, cylindrical pseudobulb, and the flowers, usually borne one or two at a time, are large in proportion to the rest of the plant and highly (sweetly) fragrant. The plant was named for Lady Lydia Ackland (1786–1856), an English orchid grower who first successfully flowered the species in Europe.

Growing near the seashore among scrubby vegetation, plants are vigorous, adaptable, and long-lived. These features, as well as the novelty of the spotted pattern, make the orchid a popular parent of many hybrids, creating progeny with compact size, vigor, and unusual coloration. Pollination is by euglossine and other large bees, which visit the flowers in search of nectar, although in this case none is available.

The flower of Lady Ackland's Orchid is large, with yellow-green, olive, or pale brownish segments with purplish-brown blotches. The three-lobed lip has the sidelobe wrapped around the column and is variable in color but usually rose purple, or lilac pink with darker purple markings.

DISTRIBUTION

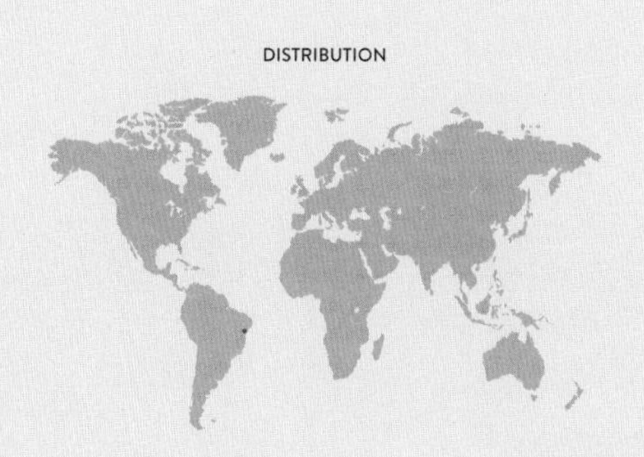

CATTLEYA COCCINEA

SCARLET CATTLEYA

LINDLEY, 1836

PLANT SIZE
4–6 × 3–5 in (10–15 × 8–13 cm),
excluding inflorescence
4–10 in (10–25 cm), just a little longer than the leaves

FLOWER SIZE
1½–3 in (3–8 cm)

SUBFAMILY ~ Epidendroideae

TRIBE AND SUBTRIBE ~ Epidendreae, Laeliinae

NATIVE RANGE ~ Southern and southeastern Brazil to Misiones province (northern Argentina)

HABITAT ~ Atlantic rain forests, at 2,130–5,480 ft (650–1,670 m)

TYPE AND PLACEMENT ~ Epiphytic on moss-covered trees or mossy rocks

CONSERVATION STATUS ~ Not assessed

FLOWERING TIME ~ April to June (fall and early winter)

Formerly placed in the genus *Sophronitis*, the Scarlet Cattleya is a beautiful miniature orchid with a relatively large flower. It forms a cluster of closely packed pseudobulbs, each bearing a single, elliptic, leathery leaf. The leaf midvein is red, and from the base of a leaf an inflorescence grows with a single, long-lasting, scarlet flower. This species has been extensively used in hybridization with other species of *Cattleya* to create bright red, large-flowered cultivars.

The red color and tubular lip of this species suggest ornithophily (bird pollination), most likely by hummingbirds. Most bees cannot see the color red, while birds with their visual acuity are certain to observe and be attracted to the red flowers. Unfortunately they will be disappointed in their expectation of a nectary treat as the deceptive flowers bear none.

The flower of the Scarlet Cattleya is brilliant red to orange (rarely yellow) with broad, spreading sepals and even larger petals. The lip is the same color but has yellow markings inside and is smaller, forward projecting, and surrounds the column.

DISTRIBUTION

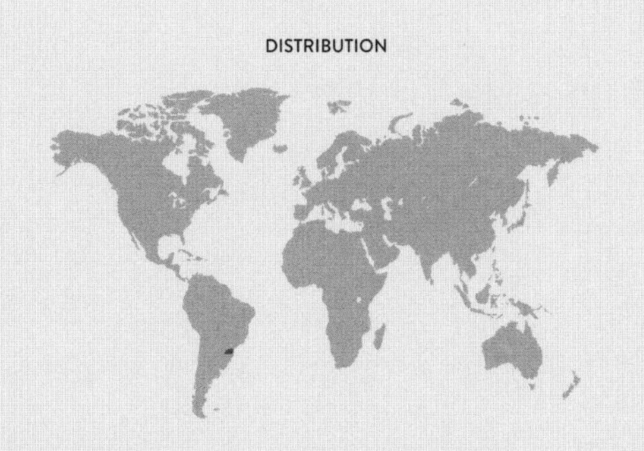

CATTLEYA MOSSIAE

EASTER ORCHID

C. PARKER EX HOOKER, 1838

PLANT SIZE
14–20 × 6–10 in (36–51 × 15–25 cm), including short, erect inflorescence, which is slightly shorter than the leaves and 8–12 in (20–30 cm) tall

FLOWER SIZE
Up to 8 in (20 cm)

SUBFAMILY ~ Epidendroideae
TRIBE AND SUBTRIBE ~ Epidendreae, Laeliinae
NATIVE RANGE ~ Venezuela
HABITAT ~ Dense montane forests, high in the canopy
TYPE AND PLACEMENT ~ Epiphytic
CONSERVATION STATUS ~ Threatened by collecting and habitat degradation
FLOWERING TIME ~ Spring, usually March to May

Native to Venezuela, the Easter Orchid has the distinction of also being the country's national flower—no mean feat in a country with many endemic beautiful species, including seven other showy species of *Cattleya*. On a single inflorescence up to five large, beautifully colored, and deliciously fragrant flowers grow, which are produced at Easter (hence the common name).

Cattleya mossiae was the second of the showy single-leafed species of the genus *Cattleya* to be discovered (*C. labiata* was the first), but *C. mossiae* was readily imported and became the earliest species available to enthusiasts in the 1830s. Information about the locality of *C. labiata* was, perhaps deliberately, misrepresented, leading to the plant being lost for 70 years before its rediscovery. Pollination of this and most other species of *Cattleya* is by nectar-seeking euglossine bees.

The flower of the Easter Orchid is large and varies in color and form. Typical forms are lavender, with narrow lanceolate sepals, broad, forward-projecting petals, and a flaring lip variably marked with darker purple and brilliant yellow spots in the throat.

DISTRIBUTION

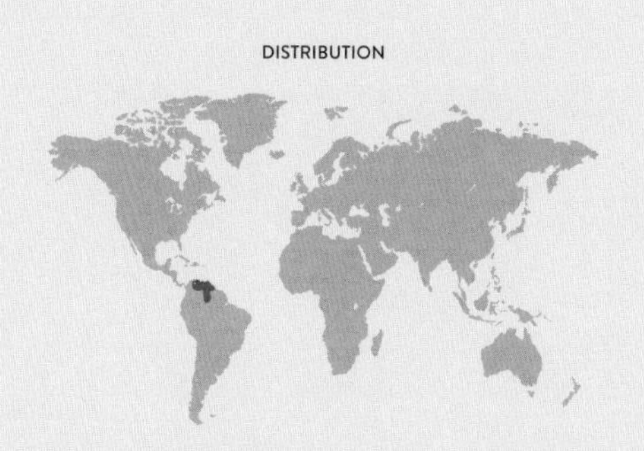

ENCYCLIA CORDIGERA

FLOWER OF THE INCARNATION

(KUNTH) DRESSLER, 1964

PLANT SIZE
16–30 × 10–18 in (41–76 × 25–46 cm), excluding erect, terminal inflorescence, 20–32 in (51–81 cm) tall

FLOWER SIZE
3½ in (9 cm)

SUBFAMILY ~ Epidendroideae

TRIBE AND SUBTRIBE ~ Epidendreae, Laeliinae

NATIVE RANGE ~ Mexico, Guatemala, Belize, El Salvador, Honduras, Nicaragua, Costa Rica, Panama, French Guiana, Surinam, Guyana, Venezuela, and Colombia

HABITAT ~ Lower elevation, seasonally dry forests, at 330–1,640 ft (100–500 m)

TYPE AND PLACEMENT ~ Epiphytic, occasionally lithophytic

CONSERVATION STATUS ~ Threatened by overcollection

FLOWERING TIME ~ Late spring to summer

The sturdy, vigorous Flower of the Incarnation is extremely variable in flower color, with many lovely forms found in nature. Flower spikes emerge from the apex of glossy, large ovoid pseudobulbs, each bearing a pair of leathery leaves. The erect spikes hold usually between six and twelve large, highly fragrant flowers well above the foliage, making a superb show. The genus name is based on the Greek word *enkyklein*, "encircle," referring to the lip, which is partially fused to and wraps around the column. The common name is simply a reference to the "incarnate" beauty of the flowers.

Encyclia cordigera is a mostly lowland species from seasonally dry areas, where it can form huge clusters of massive pseudobulbs. Large bees, especially carpenter bees, pollinate the plants, seeking nectar, which is never present.

The Flower of the Incarnation is variable in color but generally has brownish to olive petals and sepals, though purplish in some forms. The broad, flaring lip is often of a rich pink, rose, or purple, and sometimes white with a purple blotch at the center.

DISTRIBUTION

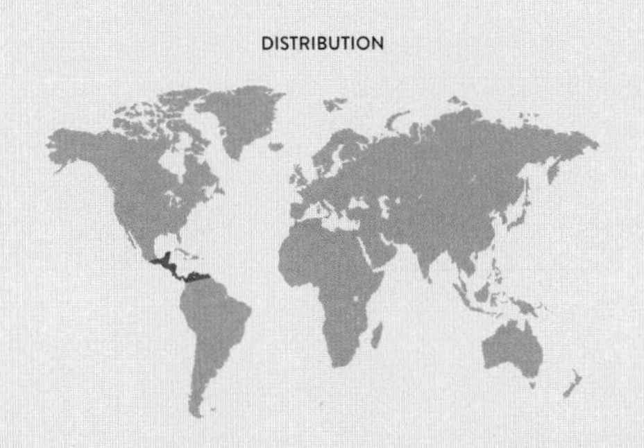

EPIDENDRUM MEDUSAE

MEDUSA-HEAD ORCHID

(REICHENBACH FILS) PFITZER, 1889

PLANT SIZE
10–16 × 4–6 in
(25–41 × 10–15 cm),
excluding short, pendent
terminal inflorescence,
which is slightly longer than
the stems, 2–3 in
(5–8 cm) long

FLOWER SIZE
2¾–4 in (7–10 cm)

SUBFAMILY ~ Epidendroideae

TRIBE AND SUBTRIBE ~ Epidendreae, Laeliinae

NATIVE RANGE ~ Ecuador

HABITAT ~ Cloud forests, at 4,920–8,200 ft (1,500–2,500 m)

TYPE AND PLACEMENT ~ Epiphytic

CONSERVATION STATUS ~ Not threatened

FLOWERING TIME ~ Mostly summer but can bloom anytime

The pendent Medusa-head Orchid resides in the cooler reaches of Ecuadorian cloud forests, its fleshy chain-like foliage hanging gracefully in virtually constant moist conditions from mossy, shady branches. The bizarre flowers, usually borne one to three at a time on short, terminal inflorescences, are variable but often have a ruby-garnet to maroon coloration with a remarkable, raggedly fringed lip, which, likened to the snake-haired head of the monster Medusa from Greek mythology, gives the plant its scientific and common names.

In this species, the lip is extensively fused to the column, as is typical for nearly all species of *Epidendrum*, and the apex of the column has a narrow slit suited to the tongue of a butterfly or moth, which are frequently found to be the pollinators of *Epidendrum*. It seems likely that these flowers are adapted for moths due to their dull color.

The flower of the Medusa-head Orchid can be variable but generally bears yellow to olive segments overlaid with a flush of maroon. The broad, oval lip is sometimes greenish but often deep dark red or maroon and ringed with a distinctive fringe.

DISTRIBUTION

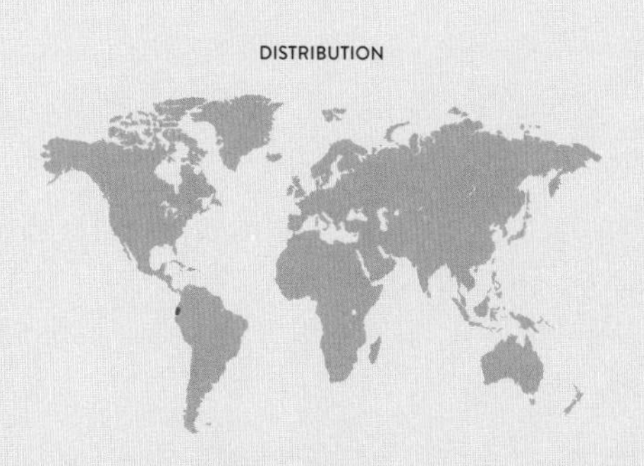

GUARIANTHE SKINNERI

SAN SEBASTIAN ORCHID

(BATEMAN) DRESSLER & W. E. HIGGINS, 2003

PLANT SIZE
15–26 × 8–15 in (38–66 × 20–38 cm), excluding the terminal inflorescence 5–8 in (13–20 cm) long

FLOWER SIZE
3½ in (9 cm)

SUBFAMILY ~ Epidendroideae

TRIBE AND SUBTRIBE ~ Epidendreae, Laeliinae

NATIVE RANGE ~ Mesoamerica, from southern Mexico to Costa Rica

HABITAT ~ Humid forests, at 650–7,545 ft (200–2,300 m)

TYPE AND PLACEMENT ~ Epiphytic on tree trunks or epilithic on granite cliffs

CONSERVATION STATUS ~ Not assessed

FLOWERING TIME ~ January to April (winter to spring, after the dry season)

The sometimes branched, creeping rhizome of the San Sebastian Orchid produces cylindrical pseudobulbs with a larger diameter near the leaves than at the base; each bears two centrally folded leaves at its apex. In the middle of the leaves, there is a large, thin floral sheath that encloses the apical bud, which in turn produces an inflorescence bearing up to 15 fragrant, brilliant purple-fuchsia flowers.

This orchid is the national flower of Costa Rica. Its genus name comes from the Costa Rican word for an epiphytic orchid (*guaria*), and the Greek *anthe*, "flower." Its local Costa Rican name is *la guaria morada*, the "darkly colored orchid," while in Guatemala, it is known as the *flor de San Sebastián*. Pollination by bees, probably carpenter (*Xylocopa*) and solitary bees (*Thygaster*), is likely, although specific reports are lacking.

The flower of the San Sebastian Orchid has three lanceolate, bright purple sepals and two much broader petals. The lip is darker purple with a white central area, tube-shaped, and folded around the column.

DISTRIBUTION

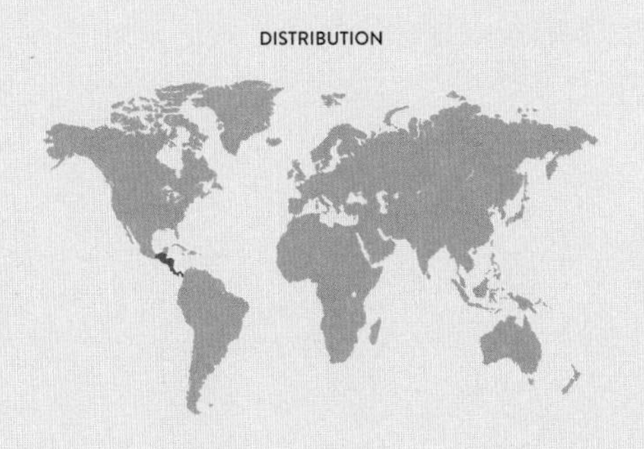

LAELIA SPECIOSA

MAYFLOWER ORCHID

(KUNTH) SCHLECHTER, 1914

PLANT SIZE
10–20 × 5–8 in (25–51 × 13–20 cm)
tall, including terminal inflorescence
5–8 in (13–20 cm) long

FLOWER SIZE
8 in (20 cm)

SUBFAMILY ~ Epidendroideae

TRIBE AND SUBTRIBE ~ Epidendreae, Laeliinae

NATIVE RANGE ~ Eastern and southern Mexico

HABITAT ~ Dry, open oak forests, at 4,600–7,875 ft (1,400–2,400 m)

TYPE AND PLACEMENT ~ Epiphytic on mossy branches

CONSERVATION STATUS ~ Threatened due to overharvesting for horticultural and religious purposes

FLOWERING TIME ~ May to August (spring to summer)

The showy Mayflower Orchid has short, round pseudobulbs topped with one or two leaves that are fleshy and tinged purple. By the end of the dry season in April, the old pseudobulb has withered and an inflorescence is produced on a newly developing pseudobulb. It bears one to four bright pink, strongly scented, large flowers, making the plant sought-after among orchid-growing enthusiasts. The plant grows slowly, sometimes maturing in 16 to 19 years, high in the mountains, where the weather is cool with occasional frosts in the winter months.

It is known locally as *flor de todos santos* ("all-saints flower"). Mexican villagers make candy from a starchy paste ground from the pseudobulbs, mixed with sugar, lemon juice, and egg white and poured into wooden molds, forming little decorative animals, fruit, and skulls for the Day of the Dead.

The flower of the Mayflower Orchid has bright purple, spreading petals and sepals, with the petals somewhat broader. The similarly colored, trilobed lip encloses the column with its lateral lobes and has a flaring, wavy-edged midlobe which has purple-red stripes on a white background.

DISTRIBUTION

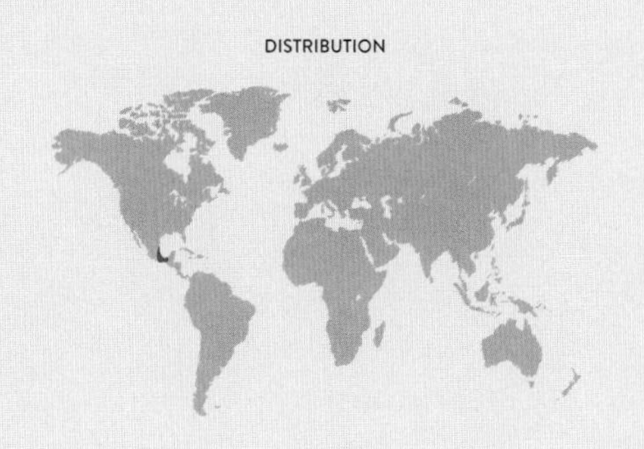

MYRMECOPHILA TIBICINIS

ANT-LOVING PIPER

(BATEMAN EX LINDLEY) ROLFE, 1917

PLANT SIZE
15–25 × 6–10 in (38–64 × 15–25 cm),
excluding terminal erect inflorescence
80–200 in (203–508 cm) tall

FLOWER SIZE
3 in (8 cm)

SUBFAMILY ~ Epidendroideae

TRIBE AND SUBTRIBE ~ Epidendreae, Laeliinae

NATIVE RANGE ~ Mexico, Guatemala, Belize, Honduras, Costa Rica, and Venezuela

HABITAT ~ Seasonally dry forests, in full sun, at 650–1,970 ft (200–600 m)

TYPE AND PLACEMENT ~ Epiphytic

CONSERVATION STATUS ~ Not threatened

FLOWERING TIME ~ July to October (summer to fall)

The giant Ant-loving Piper is a dry-forest plant that has evolved a defense against predation by animals. Its elongate pseudobulbs are hollow, with a basal opening created by an ant colony that serves as bodyguards, protecting the plant from herbivores (and naive orchid collectors). If their home is touched, the ants rush outside to defend it, a strategy reflected in the genus name, derived from the Greek words *myrmeco*, "ant," and *phila*, "loving." The species name, *tibicinis*, is Latin for a "flute-player" or "piper," as the hollow bulbs are often made into flutes by local children, hence also the common name. The ants pack older, abandoned pseudobulbs with debris, which, as it decays, provides some nutrients for the plant.

Bees, attracted by the bright colors and sweet fragrance, are the pollinators, although no reward is offered.

The flower of the Ant-loving Piper is variable in color and form but generally bears lanceolate, purplish sepals and petals with an undulate margin and darker tips. The lip is three-lobed, with the two lateral lobes bearing darker nectar guides. The midlobe is smaller and darker with a yellow central spot.

DISTRIBUTION

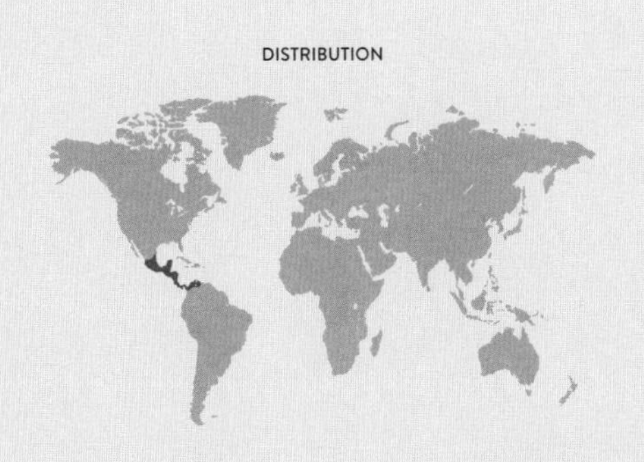

PROSTHECHEA CITRINA

TULIP ORCHID

(LEXARZA) HIGGINS, 1998

PLANT SIZE
15–25 × 6–10 in (38–64 × 15–25 cm), excluding terminal erect inflorescence 80–200 in (203–508 cm) tall

FLOWER SIZE
3 in (8 cm)

SUBFAMILY ~ Epidendroideae

TRIBE AND SUBTRIBE ~ Epidendreae, Laeliinae

NATIVE RANGE ~ Southern Mexico

HABITAT ~ Mid-elevation seasonally dry forests

TYPE AND PLACEMENT ~ Epiphytic

CONSERVATION STATUS ~ Threatened by overcollecting

FLOWERING TIME ~ Winter to spring

Like so many other large-flowered, highly ornamental New World species, the Tulip Orchid was imported in the thousands to European greenhouses (then called "stove houses"), where most perished. The common name in Mexico is *lemoncito* (little lemon), and bunches of cut flower stems are sold along the roadsides by children.

The pendent plants have large oblong-ovoid pseudobulbs and usually a pair of fleshy leaves that have an unusual glaucous sheen. The citrus-scented (hence the species name *citrina*) flowers hang from the apex of the bulbs on short stems and are pollinated by nectar-seeking euglossine bees. The plants are considered locally valuable and have been overcollected in their native range as ornamental plants, as a painkilling medicine to treat wounds, and as a source of mucilage (glue) for repairing wooden objects.

The flower of the Tulip Orchid is typically brilliant yellow (sometimes greenish-yellow), with a large, tubular lip and a white margin. There are often white and greenish nectar guides in the throat of the lip.

DISTRIBUTION

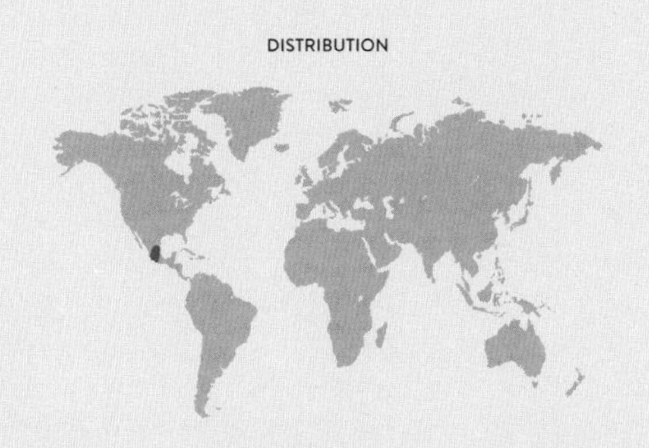

PROSTHECHEA COCHLEATA

COCKLESHELL ORCHID

(LINNAEUS) W. E. HIGGINS, 1998

PLANT SIZE
10–20 × 7–10 in (25–51 × 18–25 cm), excluding erect terminal inflorescence, which grows 8–15 in (20–38 cm) longer than the leaves

FLOWER SIZE
3 in (8 cm)

SUBFAMILY ~ Epidendroideae

TRIBE AND SUBTRIBE ~ Epidendreae, Laeliinae

NATIVE RANGE ~ Caribbean region (including Florida) and Mexico to northern South America

HABITAT ~ Tropical evergreen and deciduous oak forests, from sea level to 6,200 ft (1,900 m)

TYPE AND PLACEMENT ~ Epiphytic

CONSERVATION STATUS ~ Apparently secure due to its frequency and widespread distribution, so not of conservation concern except in Florida where it is considered endangered

FLOWERING TIME ~ Throughout the year

The Cockleshell Orchid has smooth, ovoid to elliptical, slightly flattened pseudobulbs that are enveloped basally by overlapping, dry sheaths and topped with two or three elliptic-lanceolate leaves. From the top of a mature pseudobulb, an upright inflorescence is produced that holds up to 20 upside-down—lip uppermost—scentless flowers that open over a long period of up to six months. The genus name comes from the Greek word *prostheke*, "appendage," referring to the short, pointed growth on the back of the anther.

Prosthechea cochleata is the national flower of Belize, where it is known as the Black Orchid (*orquídea negra*). The plant is fairly commonly cultivated for its unusual, long-lasting flowers, which are pollinated in nature by wasps, although they receive no reward. In Central America, mucilage is extracted from the pseudobulbs and used as glue for repairing wooden objects.

The flower of the Cockleshell Orchid has twisted and downward-pointing, greenish-yellow sepals and petals, with an upright, hoodlike, yellow lip that has bold stripes of reddish-purple fusing solidly at the margins. The column is fat and has some purple spots basally.

DISTRIBUTION

RHYNCHOLAELIA DIGBYANA

QUEEN OF THE NIGHT

(LINDLEY) SCHLECHTER, 1918

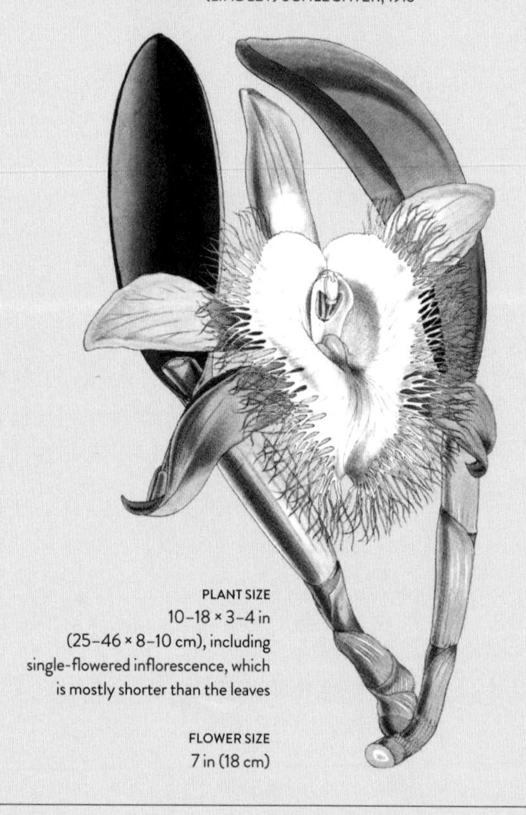

PLANT SIZE
10–18 × 3–4 in
(25–46 × 8–10 cm), including
single-flowered inflorescence, which
is mostly shorter than the leaves

FLOWER SIZE
7 in (18 cm)

SUBFAMILY ~ Epidendroideae

TRIBE AND SUBTRIBE ~ Epidendreae, Laeliinae

NATIVE RANGE ~ Southeastern Mexico to northern Honduras and Guatemala

HABITAT ~ Sunny places among bushes and thorny acacia in dry forests on limestone, at sea level to 1,640 ft (500 m)

TYPE AND PLACEMENT ~ Epiphytic

CONSERVATION STATUS ~ Not formally assessed

FLOWERING TIME ~ May to August (summer)

The flattened, closely spaced, elongate pseudobulbs of the Queen of the Night have a single elliptic, upright, fleshy leaf that is covered in a gray, dustlike substance. From the top of the pseudobulb, an upright inflorescence is formed, bearing a single flower, with a long stalk into which a long nectar cavity is embedded. The genus name comes from the Greek word *rhynchos*, "beak"—referring to the long stalk of the flowers—and its floral similarity to the genus *Laelia* (a related group in the same subtribe).

Pollination is by night-flying moths that are initially attracted by the pervasive lemon scent of the flowers. The nocturnal nature of these large blooms is responsible for the plant's common name, while the species name is in honor of a Mr. Digby, an English orchid enthusiast of the period when this species was first described in 1846.

The flower of the Queen of the Night has green, spreading, narrowly lanceolate sepals and petals, which are wider. The greenish-white lip surrounds the column and has a long-fringed margin and a raised callus creating a tube to the nectar cavity.

DISTRIBUTION

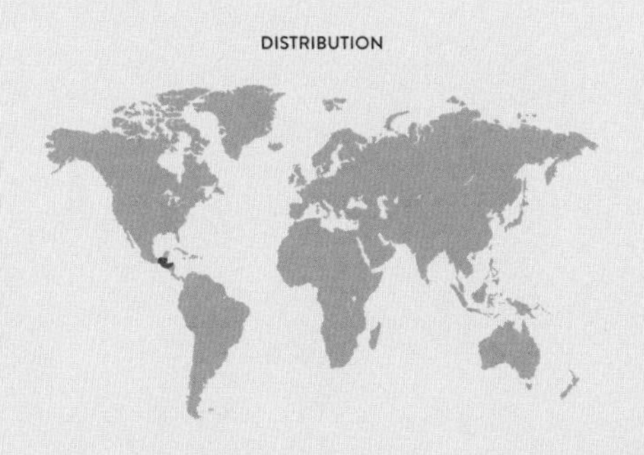

DRACULA SIMIA

MONKEY-FACE ORCHID

(LUER) LUER, 1978

PLANT SIZE
7–10 × 1–2 in (18–25 × 2.5–5 cm),
excluding pendent inflorescence
5–8 in (13–20 cm) long

FLOWER SIZE
6 in (15 cm)

SUBFAMILY ~ Epidendroideae

TRIBE AND SUBTRIBE ~ Epidendreae, Pleurothallidinae

NATIVE RANGE ~ Southeastern Ecuador

HABITAT ~ Cloud forests

TYPE AND PLACEMENT ~ Epiphytic

CONSERVATION STATUS ~ Not threatened

FLOWERING TIME ~ Spring, fall, and winter

Examined up close, the center of each pendent down-facing flower reveals what appears to be a monkey's face, from which feature both the species name *simia* (from the Latin word for "monkey") and the common name are derived. Species of the genus *Dracula* are largely wet loving, cool cloud forest dwellers with leaves that generally resemble members of *Masdevallia*. This large genus historically included *Dracula* species, and the two are still genetically compatible (and make artificial hybrids).

The long-tailed flowers are produced successively and hang below the foliage. Blooms tend to be warty and hairy with a rounded lip that bears structures reminiscent of the gills on the underside of a mushroom. Although the flowers remind us of a monkey's face, to a pollinating fungus gnat the appearance of this orchid and its fungus-like fragrance suggest that it is a good site to lay its eggs.

The flower of the Monkey-face Orchid consists of three prominent, reddish-brown sepals, each with a long tail and a sparsely haired surface. The petals are small and darkly colored. The lip is movable, white to cream, with structures similar to the gills underneath a mushroom cap.

DISTRIBUTION

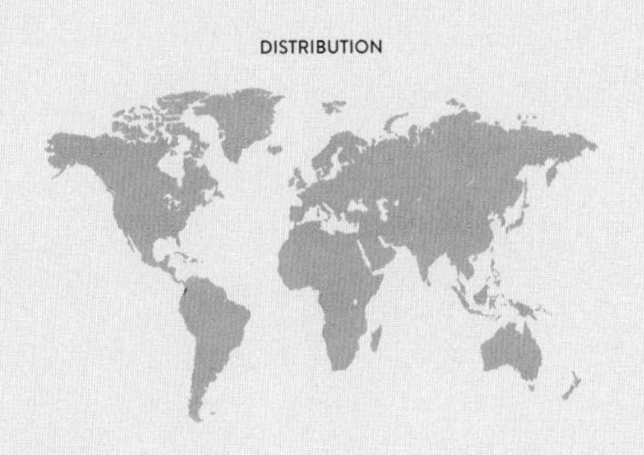

DRACULA VAMPIRA

VAMPIRE DRAGON

(LUER) LUER, 1978

PLANT SIZE
8–12 × 1–2 in (20–30 × 2.5–5 cm)

FLOWER SIZE
7 in (18 cm)

SUBFAMILY ~ Epidendroideae

TRIBE AND SUBTRIBE ~ Epidendreae, Pleurothallidinae

NATIVE RANGE ~ Western Ecuador

HABITAT ~ Cloud forests, at 5,900–7,200 ft (1,800–2,200 m)

TYPE AND PLACEMENT ~ Epiphytic

CONSERVATION STATUS ~ Threatened by poaching and collection for horticulture

FLOWERING TIME ~ Throughout the year

The Vampire Dragon has one of the most evocative names in the orchid family, conjuring up vivid images from horror movies. The genus name is derived from Medieval Latin for "little dragon," *draco, -ula*. Another source of the name is from Vlad III Dracula (1431–76), ruler of Wallachia (modern Romania), whose father was a member of the Order of the Dragon (*Dracul* in Romanian). The association with a vampire derives from Bram Stoker's Gothic novel *Dracula* (1897). The sinister-looking flowers with their bold, nearly black striping are suspended well away from the leaves on arching to pendent stems, contributing to their eerie appearance.

The species of *Dracula* all appear to be pollinated by fungus-eating gnats that are attracted to the lip, which is shaped like a mushroom, complete with "gills" and an appropriate odor. All of this impressive display is artifice.

The flower of the Vampire Dragon consists of three creamy or pale green sepals prominently overlaid with black-brown stripes, each sepal tipped with long tails. The petals are short and pale and flank the column. The lip is ladle-shaped and pinkish-cream with ridges radiating from the center.

DISTRIBUTION

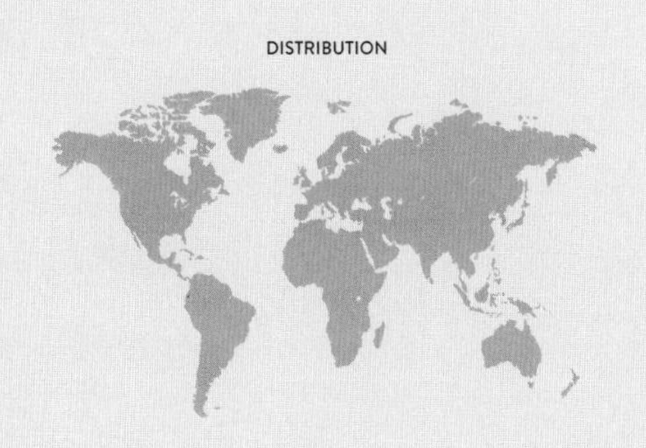

MASDEVALLIA VEITCHIANA

VEITCH'S MARVEL

REICHENBACH FILS, 1868

PLANT SIZE
6–10 × ¾–1¼ in (15–25 × 1.9–3.2 cm), excluding the erect, single-flowered inflorescence, 10–20 in (25–51 cm) tall

FLOWER SIZE
8 in (20 cm)

SUBFAMILY ~ Epidendroideae

TRIBE AND SUBTRIBE ~ Epidendreae, Pleurothallidinae

NATIVE RANGE ~ Peru, only known from Machu Picchu

HABITAT ~ Cloud forests, open rocky sites, at 6,600–13,100 ft (2,000–4,000 m)

TYPE AND PLACEMENT ~ Mostly terrestrial, sometimes lithophytic, rarely epiphytic

CONSERVATION STATUS ~ Threatened by collection for horticulture

FLOWERING TIME ~ September to December (spring and early summer)

A renowned species from the area surrounding the archaeological site at Machu Picchu, the stunning Veitch's Marvel bears a shockingly vibrant flower produced on sturdy erect stems and held well above the foliage. Often growing in full sun, the leaves are protected from sunburn by surrounding grasses. The flower color sometimes appears uneven or asymmetrical because iridescent purple hairs cover the blooms, creating a dazzling surface sheen.

Pollination of *Masdevallia veitchiana* has never been studied. However, at the elevations at which the plant grows, the pollinator was assumed to be a hummingbird, mistakenly believing nectar to be present. Recent research indicates that fungus gnats and fruit flies are actually the most likely pollination partners.

The flower of Veitch's Marvel has brilliant orange sepals fused at their bases to produce an elongate, triangular flower outline. The surface is covered with light-catching purple hairs, often producing a shimmering gleam. Petals and lip are darker and much reduced, and form a tube around the column.

DISTRIBUTION

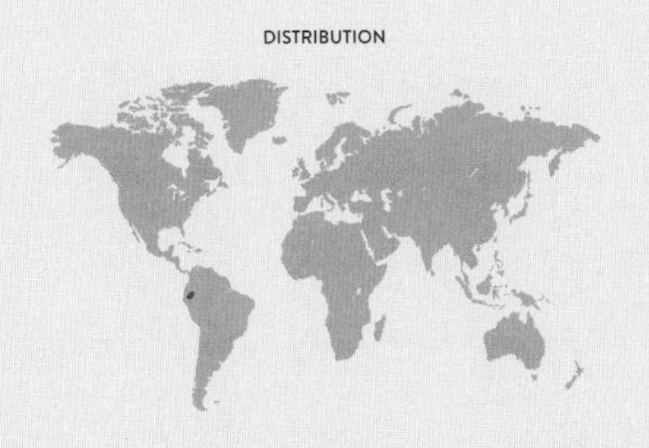

BULBOPHYLLUM GRANDIFLORUM

FOUL GIANT

BLUME, 1849

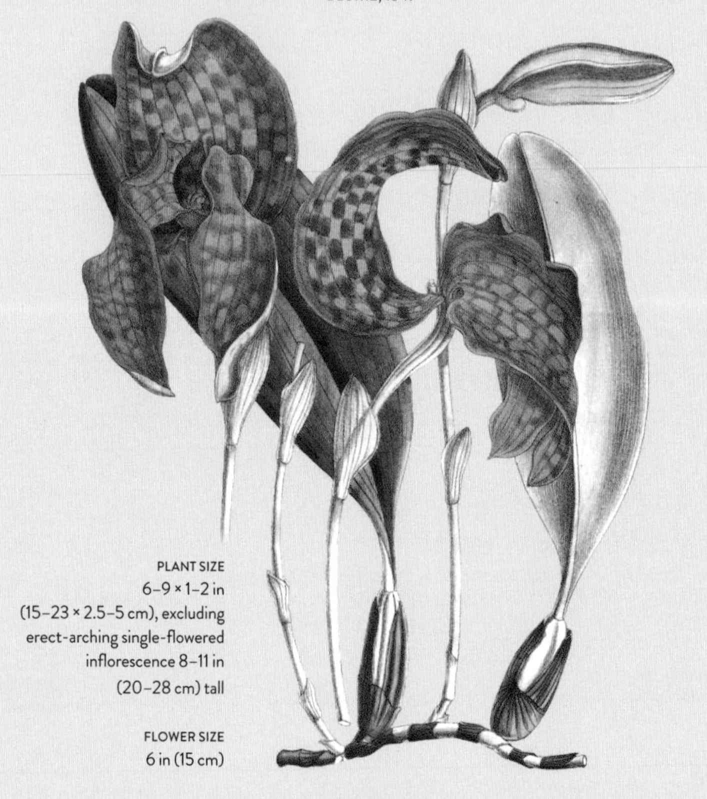

PLANT SIZE
6–9 × 1–2 in
(15–23 × 2.5–5 cm), excluding
erect-arching single-flowered
inflorescence 8–11 in
(20–28 cm) tall

FLOWER SIZE
6 in (15 cm)

SUBFAMILY ~ Epidendroideae

TRIBE AND SUBTRIBE ~ Malaxideae, Dendrobiinae

NATIVE RANGE ~ Eastern Indonesia (Sulawesi, Maluku Islands), New Guinea, and Solomon Islands

HABITAT ~ Rain forests, at 650–2,626 ft (200–800 m)

TYPE AND PLACEMENT ~ Epiphytic on lower limbs or trunks

CONSERVATION STATUS ~ Not assessed

FLOWERING TIME ~ April to May (fall)

The Foul Giant is one of the largest-flowered species in the largest orchid genus (1,900 species)—but it is also one of the most offensive. Its egg-shaped, angular pseudobulbs are borne a short distance apart and topped by a single oblong leaf. An inflorescence with two to three large sheathing bracts bears a single flower.

Pollination has not been studied in detail, but it is clear from the foul smell of the flowers that the plant is attracting a fly looking for a site to deposit its eggs. The large sepals probably serve as landing platforms, allowing the flies to wander about, during which activity they climb onto the hinged lip, pass its balance point, and are thrown into the column. As they extricate themselves, pollinia are deposited on their bodies.

The flower of the Foul Giant is dominated by three large, cream to tan sepals, sometimes with reddish-purple spots, but always with translucent spots and the dorsal sepal bending forward to cover the flower. The petals are highly reduced and green, and the lip is small and white, all with purple spots.

DISTRIBUTION

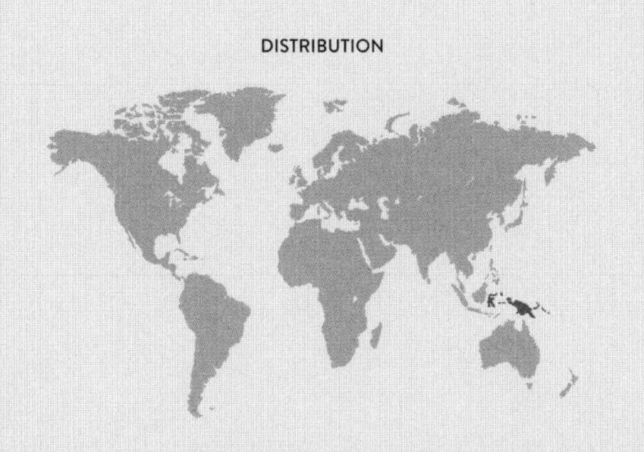

DENDROBIUM CHRYSOTOXUM

FRIED-EGG ORCHID

LINDLEY, 1847

PLANT SIZE
8–14 × 6–8 in (20–36 × 15–20 cm), including arching-pendent inflorescence 8–12 in (20–30 cm) long

FLOWER SIZE
1½–2 in (4–5 cm)

SUBFAMILY ~ Epidendroideae

TRIBE AND SUBTRIBE ~ Malaxideae, Dendrobiinae

NATIVE RANGE ~ Myanmar, Laos, Thailand, Vietnam, China, eastern Himalayas, Bangladesh, and Assam state (India)

HABITAT ~ Sea level to 1,300 ft (400 m)

TYPE AND PLACEMENT ~ Epiphytic

CONSERVATION STATUS ~ Threatened due to overcollection for use in herbal medicine

FLOWERING TIME ~ Late winter to early spring

An important medicinal plant in Southeast Asia, the Fried-egg Orchid bears copious (20 or more on a spike), honey-scented blooms. The flowers are collected and dried to produce a delicious medicinal tea said to induce peaceful, dreamless sleep. The leaves are used to treat a range of ailments, especially those associated with diabetes. The plants originate in monsoonal climates with extreme spring and summer seasonal rainfall, and this wet-dry cycle heavily influences their growth and flowering patterns.

The flowers have an exceptionally lacerate margin and appear near the apex of tall, cylindrical, slightly angled, cane-like pseudobulbs, usually in a glorious flush. Unfortunately, the splendid show lasts only between seven and ten days. The flowering coincides with the spring water-splashing festival of the Buddhist Dai people of Yunnan, China, who decorate the roofs of their houses with this orchid.

The flower of the Fried-egg Orchid can be variable in color but is generally brilliant yellow orange with a flat form and waxy sepals and petals. Some plants can have a darker orange to reddish-brown spot in the center of the lip.

DISTRIBUTION

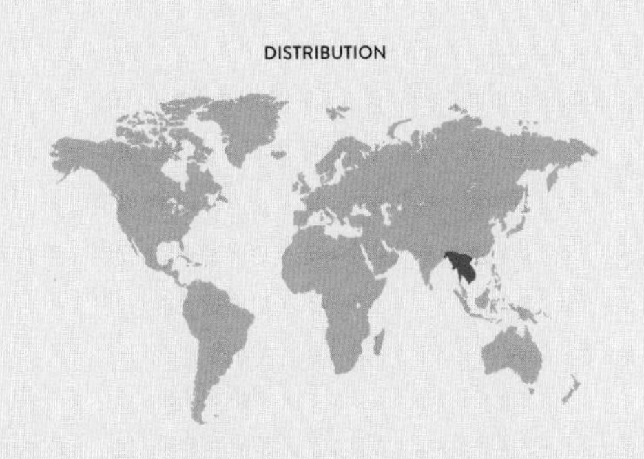

DENDROBIUM VICTORIAE-REGINAE

QUEEN VICTORIA BLUE

LOHER, 1897

PLANT SIZE
10–16 × 6–10 in (25–41 × 15–25 cm),
including inflorescence
1–2 in (2.5–5 cm) long, which grows
near the end of the pseudobulb

FLOWER SIZE
1½ in (3.8 cm)

SUBFAMILY ~ Epidendroideae

TRIBE AND SUBTRIBE ~ Malaxideae, Dendrobiinae

NATIVE RANGE ~ Philippines

HABITAT ~ Mossy, wet, cool oak forests with rhododendrons, at 4,300–8,200 ft (1,300–2,500 m)

TYPE AND PLACEMENT ~ Epiphytic

CONSERVATION STATUS ~ Not assessed

FLOWERING TIME ~ April to May (spring), but in flower almost continuously

The Queen Victoria Blue produces cane-like, longitudinally grooved pseudobulbs with papery bracts and lanceolate-ovate leaves along their length. Inflorescences can appear from the middle to the end of the often-pendent pseudobulb, carrying between two and five flowers. The genus name, from the Greek for "tree," *dendron*, and "life," *bios*, refers to the cool mossy forest habitat of the plants. The species name honors Queen Victoria, who was nearing the end of her long reign at the time the species was discovered.

Although pollination has not been studied, the shape of the flowers, their color, and their lack of scent could indicate that they are pollinated by honeyeaters or other birds, as has been found for other species of this *Dendrobium* group. The blue color of the blooms is unusual among orchids and could be mimicking flowers of *Rhododendron* species.

The flower of the Queen Victoria Blue has similar, oblanceolate, blue to lavender-purple sepals and petals with white bases and darker veins. The lip is unlobed and spoon-shaped, and also has a white base with blue stripes. The column is white and winged.

DISTRIBUTION

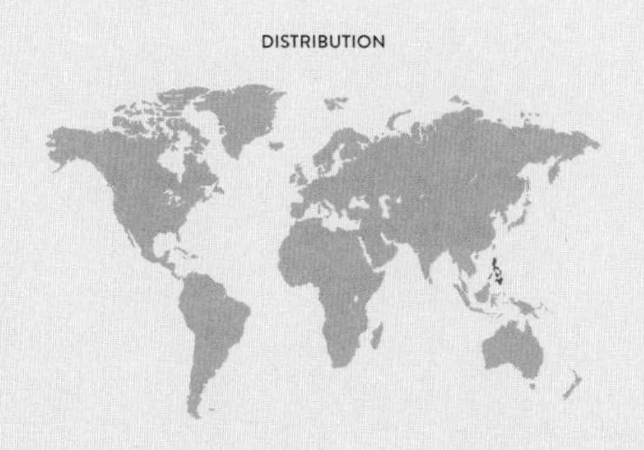

ELLEANTHUS CARAVATA

YELLOW CROWN ORCHID

(AUBLET) REICHENBACH FILS, 1881

PLANT SIZE
14–25 × 8–10 in
(36–64 × 20–25 cm),
including terminal
inflorescence

FLOWER SIZE
⅝ in (1.5 cm)

SUBFAMILY ~ Epidendroideae

TRIBE ~ Sobralieae

NATIVE RANGE ~ French Guiana, Surinam, Guyana, Venezuela, Ecuador, Brazil, and the Windward Islands

HABITAT ~ Seasonally wet forests, at 650–4,920 ft (200–1,500 m)

TYPE AND PLACEMENT ~ Epiphytic

CONSERVATION STATUS ~ Not threatened

FLOWERING TIME ~ May to September (late spring through fall)

The epiphytic Yellow Crown Orchid is one of the prettiest of a genus with many eye-catchingly colorful inflorescences. It is more compact than many of its sister species, with a comparatively large and showy terminal inflorescence, bearing small, tubular, yellow blooms subtended by contrasting, reddish-purple bracts. Like all *Elleanthus* species, the plant, which produces leafy stems, has no pseudobulbs. The species name is from the Greek *karabos*, a type of boat, alluding to the shape of the bracts. The genus name is from the Greek *Elle*, Helen (of Troy legend), and *anthos*, "flower."

The flowers and especially their colorful bracts make a long-lasting display that acts as a beacon for the hummingbirds that pollinate this and nearly all species of *Elleanthus*. The strikingly different pairing of colors is found in many bird-pollinated flowers.

The flower of the Yellow Crown Orchid is brilliant yellow with a purple tipped column. The sepals and petals are lanceolate and form a tube around the column together with the lip, which has a frilly margin and is positioned on the upper side of the flower away from the bract.

DISTRIBUTION

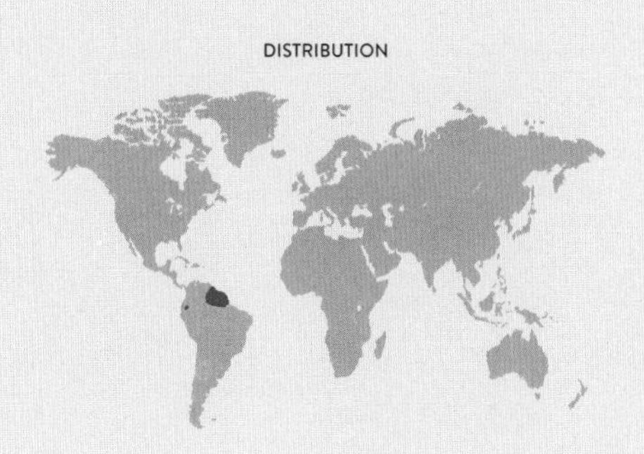

SOBRALIA MACRANTHA

LARGE PURPLE DAY-ORCHID

LINDLEY, 1838

PLANT SIZE
48–60 × 10–18 in (122–152 × 25–46 cm), including inflorescence, which is apical and short

FLOWER SIZE
8–10 in (20–25 cm)

SUBFAMILY ~ Epidendroideae

TRIBE ~ Sobralieae

NATIVE RANGE ~ Western Mexico to Costa Rica

HABITAT ~ Mostly shaded sites, especially steep slopes (including roadsides) with at least seasonally wet conditions

TYPE AND PLACEMENT ~ Terrestrial

CONSERVATION STATUS ~ Widespread and common, but not assessed

FLOWERING TIME ~ March to October

The species epithet *macrantha*, from the Greek for "large-flowered," refers to the exceptional size of its bloom. *Sobralia* species are tall, bamboo-like plants with short-lived flowers, often lasting only a few hours—an important aspect of their ecology. Tending to flower in flushes triggered by weather events, different species will often bloom en masse but on different days, thus avoiding hybridization. Pollination is by euglossine bees (males and females) and bumblebees (*Bombus*). Production of nectar has not been documented, but the high rates of visitation observed suggest that it must be present.

In recent years, the Large Purple Day-orchid, with its many color forms, has been hybridized with other *Sobralia* species. However, the large size of the plants and the short life of the flowers limit their horticultural appeal.

The flower of the Large Purple Day-orchid is typically a rich dark purple with a yellow and white throat. Other color forms are also common, including white with a yellow throat and white with a pink lip and a yellow throat.

DISTRIBUTION

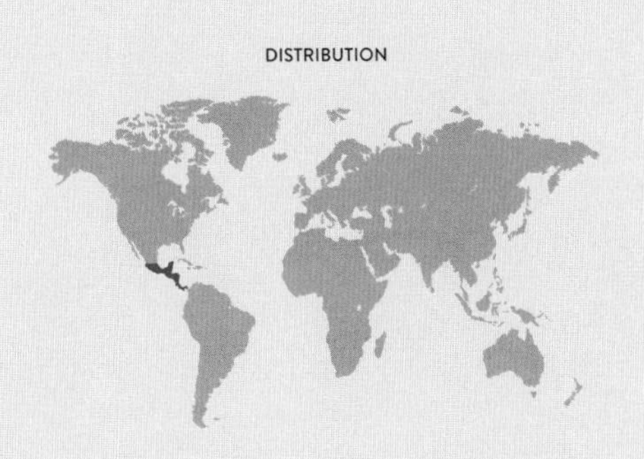

DIMORPHORCHIS LOWII

DIMORPHIC TIGER ORCHID

(LINDLEY) ROLFE, 1919

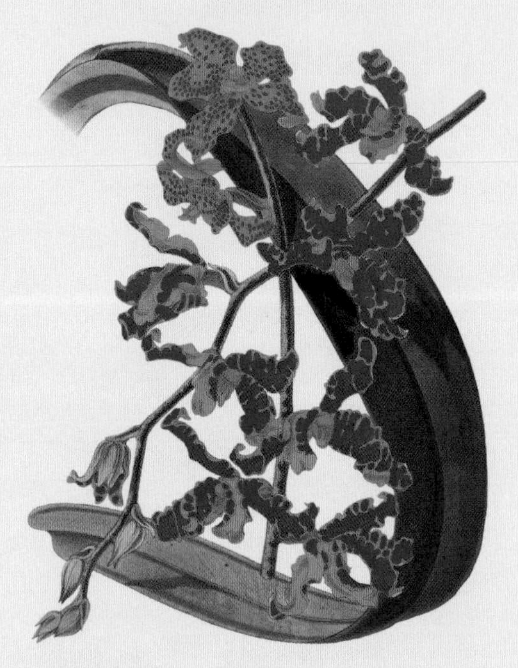

PLANT SIZE
25–50 × 20–36 in (64–127 × 51–91 cm), excluding inflorescence, which is pendent and up to 12 ft (3.65 m) long

FLOWER SIZE
3 in (7.5 cm)

SUBFAMILY ~ Epidendroideae

TRIBE AND SUBTRIBE ~ Vandeae, Aeridinae

NATIVE RANGE ~ Borneo, especially state of Sarawak

HABITAT ~ Lower montane forests, ravines, gullies, up to 5,900 ft (1,800 m)

TYPE AND PLACEMENT ~ Epiphytic, usually in tall trees overhanging water

CONSERVATION STATUS ~ Not assessed

FLOWERING TIME ~ During or after the monsoon (fall to early winter)

The robust, sometimes massive, Dimorphic Tiger Orchid grows an erect to slightly arching elongate stem with thick, fleshy, aerial roots near its base. Its lanceolate, folded leaves are unevenly lobed at the tip and sheathing at their bases. From a leaf axil it produces a limp, pendent, long inflorescence with many flowers.

Each inflorescence has flowers of two types: at the base of the inflorescence are two fragrant yellow flowers with small red dots, whereas the other flowers are creamy white with large red blotches and nearly scentless. Both types of flowers appear to have a complete column with pollinia and a functioning stigma, so the floral dimorphism does not appear to be associated with sexual dimorphism. Given the flower morphology, especially the lack of a nectar spur, the species should be bee-pollinated, but this has never been reported.

The flower of the Dimorphic Tiger Orchid has spreading, wavy sepals and two forward-pointing, smaller petals. The lip is cup-shaped and forms an opening to the column.

DISTRIBUTION

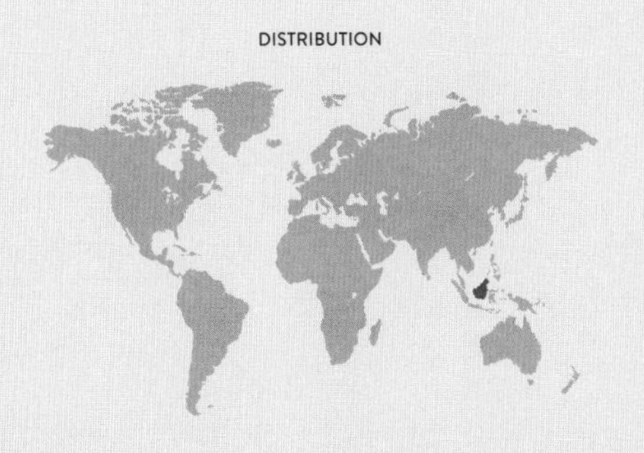

PHALAENOPSIS AMABILIS

WHITE MOTH ORCHID

(LINNAEUS) BLUME, 1825

PLANT SIZE
3–5 × 8–15 in (8–13 × 20–38 cm), excluding erect to arching inflorescence 20–35 in (51–89 cm) long

FLOWER SIZE
3 in (8 cm)

SUBFAMILY ~ Epidendroideae

TRIBE AND SUBTRIBE ~ Vandeae, Aeridinae

NATIVE RANGE ~ Borneo and the Philippines to Queensland (Australia)

HABITAT ~ Wet evergreen forests, often overhanging streams and swamps, from sea level to 1,970 ft (600 m)

TYPE AND PLACEMENT ~ Epiphytic

CONSERVATION STATUS ~ Not assessed, but threatened in some parts of its range by collection for horticulture

FLOWERING TIME ~ April to July (spring to summer, winter in Australia)

The White Moth Orchid has between four and six thick, ovate-elliptic leaves arranged in two ranks on a short stem. The inflorescence emerges from near the base of the stem and can carry up to 40 large, sweetly fragrant flowers. The species was one of the few tropical Asian orchids known to Linnaeus, and he placed it with all other epiphytic orchids in the genus *Epidendrum*, on the simple basis that they grew on trees. The genus name is Greek for "like a moth" (*phalaina*, "moth"), and its member species, along with many horticultural hybrids, are collectively known as moth orchids.

In spite of its mothlike shape, the species is pollinated by carpenter bees (genus *Xylocopa*). There is no reward, but the combination of the large showy flowers with nectar guides (spots and stripes) and sweet fragrance is enough to attract these large insects.

The flower of the White Moth Orchid has outstretched, elliptical sepals and broad, axe-shaped petals. The sidelobes of the trilobed lip encircle the column, and the yellow, red-spotted or striped midlobe has four lobes or appendages, the apical pair hairlike. A prominent callus sits in the middle of the three lobes.

DISTRIBUTION

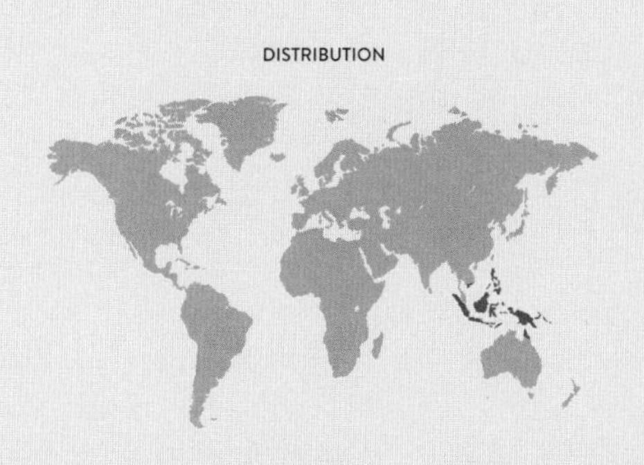

PHALAENOPSIS JAPONICA

NAGO ORCHID

(REICHENBACH FILS) KOCYAN & SCHUITEMAN, 2014

PLANT SIZE
3–4 × 5–6 in
(8–10 × 13–15 cm), including
arching-pendent inflorescence
5–8 in (13–20 cm) long

FLOWER SIZE
1 in (2.5 cm)

SUBFAMILY ~ Epidendroideae

TRIBE AND SUBTRIBE ~ Vandeae, Aeridinae

NATIVE RANGE ~ Southern China, southern Korea, and southern Japan (including Okinawa Island)

HABITAT ~ Tree trunks in open forests or along valleys, at 1,640–4,600 ft (500–1,400 m)

TYPE AND PLACEMENT ~ Epiphytic

CONSERVATION STATUS ~ Not assessed, but potentially endangered due to collection for cultivation throughout its range

FLOWERING TIME ~ May to August (spring to summer)

The Nago Orchid is a small species that produces a short stem clothed with closely spaced elliptic leaves that basally wrap around the stem. In many cases, two or more inflorescences are produced at the same time, and when each of them holds between six and ten flowers the mass of blooms can almost completely obscure the leaves. The common name comes from the city on Okinawa island, where this species was originally collected. It was formerly considered a member of genus *Sedirea* until DNA studies showed it to be a member of *Phalaenopsis*.

Pollination is by small bees that are attracted by the highly fragrant, orange flower-scented blooms. The colorful flowers appear to the bees to contain nectar, but no reward is offered—another example of deceit pollination.

The flower of the Nago Orchid has white sepals and petals that project slightly forward. The lateral sepals often have deep purple-red bars. The lip is white with lavender-purple marking and a large, flaring, apical lobe with two short lateral lobes that flank the opening into the spur.

DISTRIBUTION

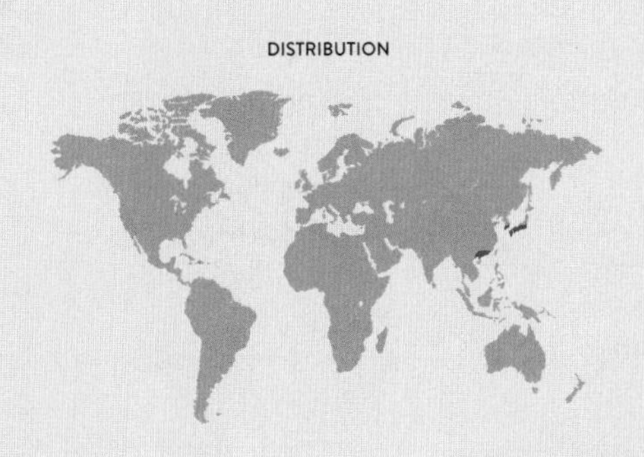

RHYNCHOSTYLIS COELESTIS

BLUE FOXTAIL ORCHID

(REICHENBACH FILS) A. H. KENT, 1891

PLANT SIZE
8–12 × 10–18 in (20–30 × 25–46 cm), excluding erect lateral inflorescence 10–18 in (25–46 cm) long

FLOWER SIZE
¾ in (2 cm)

SUBFAMILY ~ Epidendroideae

TRIBE AND SUBTRIBE ~ Vandeae, Aeridinae

NATIVE RANGE ~ Southern Indochina

HABITAT ~ Semi-deciduous, dry woodland and savannas, at up to 2,300 ft (700 m)

TYPE AND PLACEMENT ~ Epiphytic

CONSERVATION STATUS ~ Not assessed

FLOWERING TIME ~ July to November (summer to fall)

The Blue Foxtail Orchid continues to grow from its apex over many years, and its upright stem bears several strap-like, fleshy leaves. From the bases of these leaves, it produces many waxy, fragrant flowers on one to several inflorescences borne simultaneously, with possible color variants including pink and white forms. The genus name is derived from Greek and refers to the beaked (*rhynchos*) column (*stylis*).

The generally blue color of the flowers makes them popular in horticulture; hybrids with *Vanda*, to which the species is closely related, are popular. Species of *Rhynchostylis*, such as *R. retusa*, are used medicinally in Nepal, India, and Sri Lanka to treat wounds and rheumatism. The dried flowers are also reportedly useful as an insect repellant.

The flower of the Blue Foxtail Orchid has spreading, white to pale blue petals and sepals, and a lip with a darker apex that forms a prominent spur at the back. In some forms, the tips of the sepals and petals have a dark splash.

DISTRIBUTION

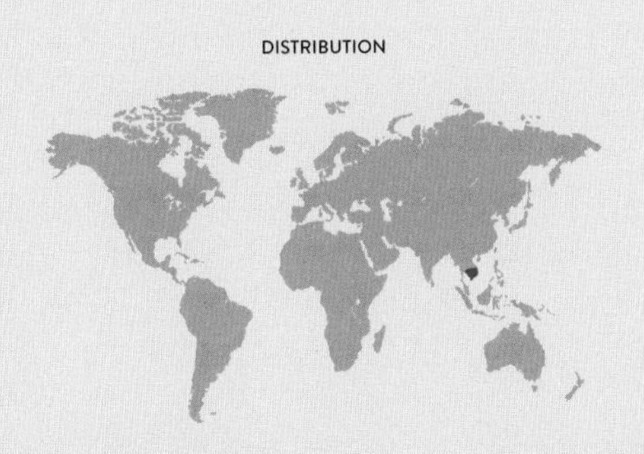

VANDA COERULEA

BLUE VANDA

GRIFFITH EX LINDLEY, 1847

PLANT SIZE
20–75 × 18–30 in
(51–191 × 46–76 cm), excluding
inflorescence, which is erect to arching,
10–30 in (25–76 cm) long

FLOWER SIZE:
4–4¾ in (10–12 cm)

SUBFAMILY ~ Epidendroideae

TRIBE AND SUBTRIBE ~ Vandeae, Aeridinae

NATIVE RANGE ~ From Assam and the Khasi Hills (India) to Yúnnan province (China), Myanmar, and northern Thailand

HABITAT ~ Dry-deciduous forests, at 2,625–5,600 ft (800–1,700 m)

TYPE AND PLACEMENT ~ Epiphytic on exposed deciduous trees, primarily dwarf oak

CONSERVATION STATUS ~ Originally listed as highly endangered, and it remains so in Assam, where it was first found, but is widespread and locally common in eastern Himalayas

FLOWERING TIME ~ September to November (fall)

When the Blue Vanda, with its stout stems, fans of folded leathery leaves, and beautifully checkered blue flowers, was first discovered by William Griffith in 1847, it caused a big stir because such a large blue orchid, offering spikes of flat long-lasting flowers, was an orchid grower's dream. The discovery resulted in several expeditions to Assam to collect this and other remarkable plants for the hothouses of Europe.

Vanda coerulea is responsible for the vibrant blues and purples of many cultivated *Vanda* hybrids. The scientific name is derived from *vandaar*, the vernacular Sanskrit name for epiphyte. Juice from the flower has been used to create eyedrops for treating glaucoma and cataracts. Laboratory research has also indicated that extracts of this blue orchid may have potential for use in antiaging skin treatments.

The flower of the Blue Vanda has clawed, broad, spreading sepals and petals, the latter often with a twisted stalk. The lip is short and three-lobed, and the column has a white cap.

DISTRIBUTION

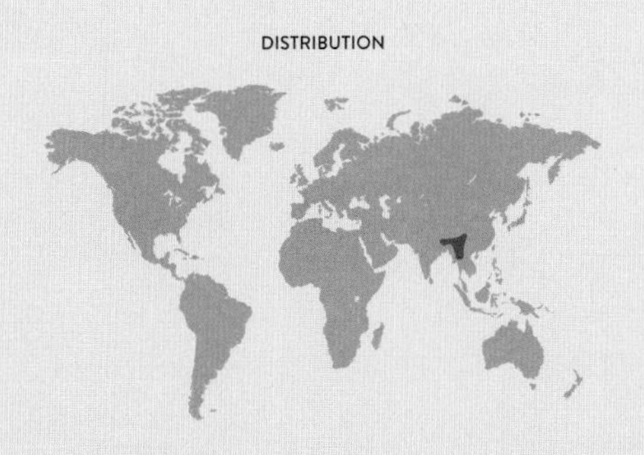

VANDA CURVIFOLIA

VERMILION BOTTLEBRUSH VANDA

(LINDLEY) L. M. GARDINER, 2012

PLANT SIZE
6–20 × 8–15 in (15–51 × 20–38 cm),
excluding lateral, erect inflorescence
5–12 in (13–30 cm) tall

FLOWER SIZE
1 in (2.5 cm)

SUBFAMILY ~ Epidendroideae

TRIBE AND SUBTRIBE ~ Vandeae, Aeridinae

NATIVE RANGE ~ Eastern Himalayas to Thailand

HABITAT ~ Semi-deciduous and deciduous dry forests, from sea level to 2,300 ft (700 m)

TYPE AND PLACEMENT ~ Epiphytic on deciduous trees

CONSERVATION STATUS ~ Not assessed

FLOWERING TIME ~ March to May (late spring to summer)

The short often basally branching, stout stems of the Vermilion Bottlebrush Vanda are covered by two rows of narrowly linear, strongly curving leaves that are two-toothed at the tip and have sheathing bases. The species produces densely flowered, upright inflorescences of 20–60 bright orange-red flowers. This species was well known in the previously recognized genus *Ascocentrum*, but is now included in a broader concept of the genus *Vanda*, which gets its name from the Sanskrit word for an epiphyte.

The flowers are scentless, have dark pollinia caps, and produce nectar in a relatively large spur lined by secretory hairs in its middle section. These traits, plus the color, are typical for bird pollination, although in this case it has not been confirmed by studies in the wild.

The flower of the Vermilion Bottlebrush Vanda is red-orange and has spreading, broad, oblanceolate sepals and petals of similar size and shape. The lip is simple, lanceolate, and curved backward. At its base, in the middle of a pair of yellow knobs, is an opening to the nectary.

DISTRIBUTION

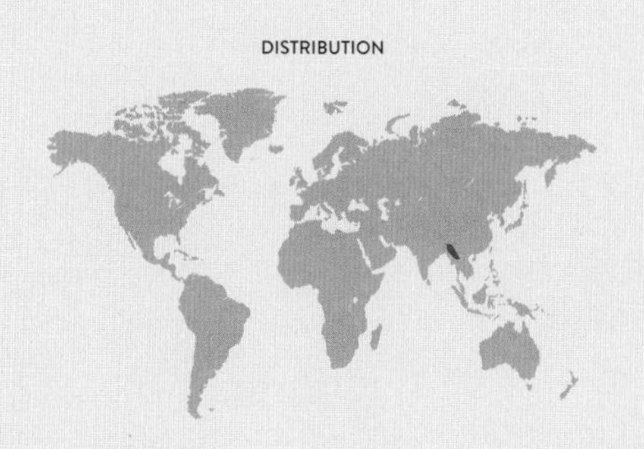

VANDA FALCATA

WIND ORCHID

(THUNBERG) BEER, 1854

PLANT SIZE
5–8 × 6–10 in (13–20 × 15–25 cm), excluding axillary, mostly erect inflorescence 2–5 in (5–13 cm) long

FLOWER SIZE
1½ in (3.8 cm)

SUBFAMILY ~ Epidendroideae

TRIBE AND SUBTRIBE ~ Vandeae, Aeridinae

NATIVE RANGE ~ China, Japan, and Korea

HABITAT ~ Deciduous forests, at 1,640–3,300 ft (500–1,000 m)

TYPE AND PLACEMENT ~ Epiphytic and lithophytic

CONSERVATION STATUS ~ Not threatened

FLOWERING TIME ~ Late spring and summer

The Wind Orchid has been treasured in East Asia for centuries as a cultivated plant. It has a revered cultural significance in its native range, despite being cultivated all over the world in recent decades. As recently as 2014, the species was treated as a member of the genus *Neofinetia*, but this small epiphyte was then transferred back to the much larger genus *Vanda*, in which it had been included in 1854. *Vanda* is the Sanskrit word for "epiphyte," and its common name is a translation of its name in Japanese, *fuuran*, reflecting the graceful form of the flower. Many hybrids between this species and other genera in subtribe Aeridinae have been produced.

In its native habitats, *V. falcata* is pollinated by hawk moths. They are attracted by the orchid's sweet scent, only produced at night, and the nectar in its long spur.

The flower of the Wind Orchid is generally white, although pale purple, pink, yellow, and green-tinted forms exist, most with an entirely white lip. Sepals and petals are narrowly lanceolate and reflexed, and the trilobed lip leads to a long, curved nectar spur.

DISTRIBUTION

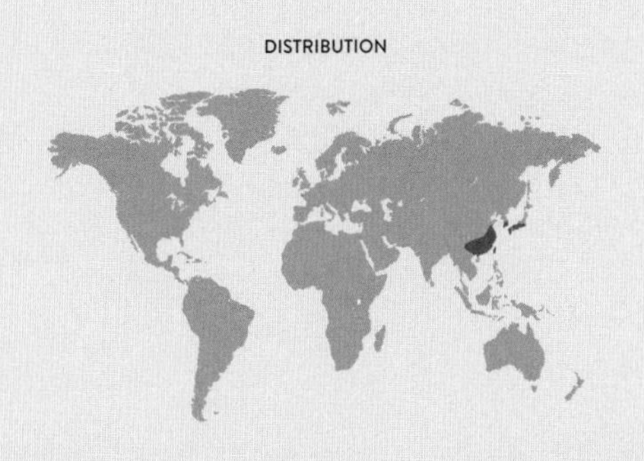

ANGRAECUM SESQUIPEDALE

COMET ORCHID

THOUARS, 1822

PLANT SIZE
18–36 × 24–40 in
(46–91 × 61–102 cm),
including arching inflorescences
15–18 in (38–46 cm) tall

FLOWER SIZE
6–8 in (15–20 cm)

SUBFAMILY ~ Epidendroideae

TRIBE AND SUBTRIBE ~ Vandeae, Angraecinae

NATIVE RANGE ~ Madagascar

HABITAT ~ Rain forests, at up to 330 ft (100 m)

TYPE AND PLACEMENT ~ Epiphytic and lithophytic

CONSERVATION STATUS ~ Threatened by poaching

FLOWERING TIME ~ Fall to spring

The remarkable Comet Orchid bears one of the most extraordinary flowers of any plant. The species epithet, from the Latin words *sesqui*, meaning "one and a half," and *pedalis*, for "foot," refers to the prodigious length of the nectary at the back of the flower. The large fan-shaped plants, with thick, coarse aerial roots, grow epiphytically on larger tree branches near sea level. The genus name comes from the Indonesian word for orchid, *anggrek*.

When Charles Darwin observed the structure of these flowers in 1862, he hypothesized the existence of a then undescribed moth species with a proboscis long enough to reach the full length of the nectar spur, up to 12 in (30 cm). Darwin was vindicated years later (1903) when such a moth was discovered in Madagascar, *Xanthopan morgani*. Actual pollination has now been observed and filmed.

The flower of the Comet Orchid is star-shaped and pure glistening white, with a triangular lip. When they first open, the flowers have a greenish tinge. The most noticeable feature is the exceptionally long nectar spur, in which nectar can often be observed in the bottom portion.

DISTRIBUTION

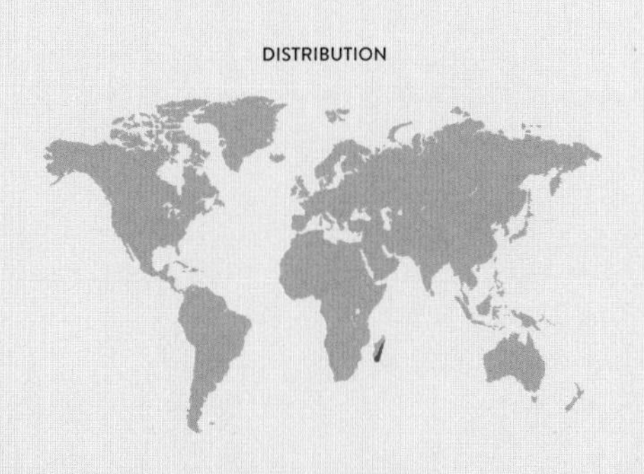

PLECTRELMINTHUS CAUDATUS

WORM ORCHID

(LINDLEY) SUMMERHAYES, 1949

PLANT SIZE
10–30 × 10–30 in (25–76 × 25–76 cm), excluding arching-pendent inflorescence 15–25 in (38–64 cm) long

FLOWER SIZE
3 in (8 cm)

SUBFAMILY ~ Epidendroideae

TRIBE AND SUBTRIBE ~ Vandeae, Angraecinae

NATIVE RANGE ~ Widespread in tropical and subtropical West and Central Africa

HABITAT ~ Low on large trees in forests, at sea level to 3,300 ft (1,000 m)

TYPE AND PLACEMENT ~ Epiphytic

CONSERVATION STATUS ~ Not threatened

FLOWERING TIME ~ November to January, but can bloom anytime

The leathery-leaved Worm Orchid has stunning, non-resupinate flowers with astonishing twisting, wormlike nectar spurs, which give rise to both its common and genus names (Greek for "spur," *plektron*, and "worm," *minthion*). The species name comes from the Latin word for "tailed," also a reference to the narrow spur.

The flowers appear alternately in two ranks, with the blooms oddly facing slightly inward rather than outward. The prodigious curling spurs and nocturnal fragrance indicate a large, long-tongued hawk moth as the likely pollinator, but it is difficult to imagine how the tongue of the moth deals with the curves in the spur. The apex of the column projects into the mouth of the nectar spur, contacting the underside of the moth's body as it approaches to insert its tongue.

The flower of the Worm Orchid has narrow, olive green to almost orange sepals and petals with pointed tips. The flowers alternate on opposite sides and have a pure white, spade-shaped lip with a long, narrow, lance-shaped midlobe pointing straight up.

DISTRIBUTION

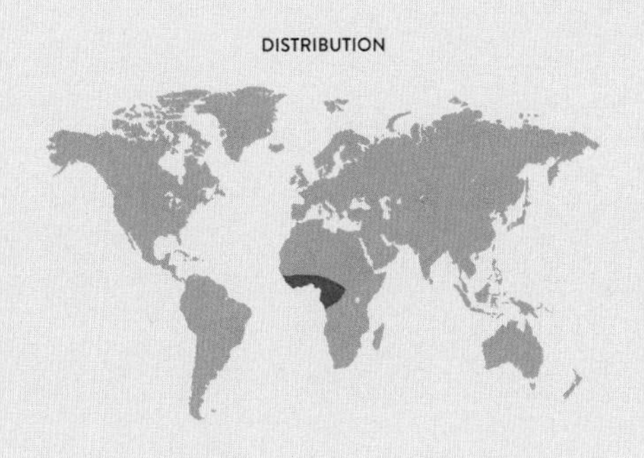

GLOSSARY

Acuminate With a pointed tip.

Adnate Fused.

Anther Pollen-bearing structure of a flower.

Apex (adj. apical) Point at the upper end of a structure.

Apomictic Producing seeds without pollination.

Axil (adj. axillary) Upper side of the point where a leaf attaches to a stem.

Basal Portion of a structure nearest to the point of attachment.

Boreal Subarctic regions.

Bract Small leaflike structure.

Buzz pollination Form of pollination in which the anthers have apical pores and are gathered together such that the tips all touch; the pollinating bee grabs these anthers and buzzes its flight muscles, making the pollen erupt from the anthers, at which point the bee gathers the pollen.

Calcareous Chalky.

Callus (pl. calli) Swollen structure on a flower, usually helping to direct the movement of the pollinator or the pollinator's tongue.

Cilia (adj. ciliate) Large hairs.

Circumboreal Occurring throughout the boreal (subarctic) zone.

Cordate Heart-shaped.

Corm Swollen stem that is dormant during the winter or dry season.

Cristate With a crest.

Dimorphic With two forms.

Ectomycorrhizal Connected by the roots to a fungus.

Epilithic Growing on rocks.

Epiphytic Growing on a tree, but not a parasite.

Falcate Curved like a sickle.

Fusiform Tapering at both ends, spindle-shaped.

Glaucous With a powdery appearance.

Globose Globe-shaped.

Inflorescence Stem that bears the flowers.

Keel Elongate raised area.

Labellum Lip, the modified petal of an orchid.

Lamella (pl. lamellae, adj. lamellate) Elongate raised area.

Lamina Flat part of a leaf (not the stem).

Lanceolate Lance-shaped.

Lithophytic Growing on rocks.

Montane Growing on mountains.

Morphology Shape and structure of a plant.

Mucilage Slime.

Mycorrhizal Referring to a beneficial relationship with fungus.

Non-resupinate With the lip uppermost, achieved by a twisting of the stem holding the flower or the flower simply flipping over on its axis.

Oblanceolate (also linear-oblanceolate) Leaf or flower part with the basal end narrower than the apical end.

Obligate Epiphyte growing only on trees.

Ovate Leaf or flower part with the broader part at the base.

Pandurate (also panduriform) Violin-shaped (thinner in the middle).

Pilose Hairy.

Plicate (of leaves) Folded, like a lady's fan.

Pollinarium Structure, consisting of the pollinia, stipe, and viscidium (see definitions below), associated with the pollen in an orchid.

Pollinium (pl. pollinia) Mass of non-powdery pollen.

Pseudobulb Swollen part of an orchid stem.

Pseudopollen Structures produced by a flower to resemble pollen.

Raceme Unbranched flower stem in which the oldest flowers are at the base.

Recurved Curved back toward the base.

Resupinate With the lip lowermost, achieved by a twisting of the stalk.

Rhizome Underground stem that grows horizontally.

Rugose Wrinkled.

Saccate Sack-shaped.

Sepal Outermost part of a flower.

Sidelobe Lobe other than the apical lobe of a flower part.

Stamen Pollen-bearing structure.

Staminode Sterile stamen.

Stipe Stalk, part of the pollinarium attaching the pollinia to viscidium.

Style Stem that bears the stigma (the structure in a flower that receives the pollen).

Substrate Material in which a plant grows.

Subtend To sit below.

Sulcate With a groove (sulcus).

Tepal Petals and sepals.

Terrestrial Growing in soil.

Tuber Swollen underground stem (like a potato).

Umbel Flower stem in which the flowers are all attached to a central point.

Understory Ground section of a forest.

Venation Collective system of veins in a leaf or floral part.

Ventral Lower or on the underside.

RESOURCES

BOOKS AND JOURNALS

Chase, M. W., M. Christenhusz, and T Mirenda. *The Book of Orchids.* Ivy Press, 2017.
A comprehensive collection of profiles of more than 600 species of orchids, from which *The Little Book of Orchids* is derived.

Chase, M. W., K. M. Cameron, J. V. Freudenstein, A. M. Pridgeon, G. Salazar, C. van den Berg, and A. Schuiteman. 'An updated classification of Orchidaceae.' *Botanical Journal of the Linnean Society* 177: 151–174 (2015).
A list of orchid genera and the number of species in each with a review of recently published phylogenetic papers on orchids.

Van der Cingel, N. H. *An Atlas of Orchid Pollination: European Orchids.* CRC Press, 2001.

Van der Cingel, N. H. *An Atlas of Orchid Pollination: America, Africa, Asia and Australasia.* Balkema, 2001.

Davy, A. and D. Gibson. Virtual issue: Charismatic Orchids. *Journal of Ecology*, 2015: www.journalofecology.org/view/0/orchidVI.html.
This is a compilation of orchid articles published in recent years in the *Journal of Ecology*; the topics covered include biological flora of the British Isles, demographic studies, mycorrhizal associations, and reproductive ecology.

Fay, M. F. and M. W. Chase. Orchid biology: from Linnaeus via Darwin to the 21st century. *Annals of Botany* 104: 359–364 (2009).
A review article from a volume of this journal dedicated to orchids, covering many areas of orchid biology.

Kull, T., J. Arditti, and S. M. Wong. *Orchid biology: reviews and perspectives* X (2009). Springer.
This is the tenth volume in a series that began in 1977, most volumes edited by Arditti; it is comprised of solicited chapters on various aspects of orchid biology, history, cultivation, and even orchids in space.

Pridgeon, A. M., P. J. Cribb, M. W. Chase, and F. N. Rasmussen. *Genera Orchidacearum, Vol. 1: General Introduction, Apostasioideae and Cypripedioideae.* Oxford University Press, 1999.

Pridgeon, A. M., P. J. Cribb, M. W. Chase, and F. N. Rasmussen. *Genera Orchidacearum, Vol. 2: Orchidoideae (Part one).* Oxford University Press, 2001.

Pridgeon, A. M., P. J. Cribb, M. W. Chase, and F. N. Rasmussen. *Genera Orchidacearum, Vol. 3: Orchidoideae (Part two) and Vanilloideae.* Oxford University Press, 2003.

Pridgeon, A. M., P. J. Cribb, M. W. Chase, and F. N. Rasmussen. *Genera Orchidacearum, Vol. 4: Epidendroideae (Part one)*. Oxford University Press, 2005.

Pridgeon, A. M., P. J. Cribb, M. W. Chase, and F. N. Rasmussen. *Genera Orchidacearum, Vol. 5: Epidendroideae (Part two)*. Oxford University Press, 2009.

Pridgeon, A. M., P. J. Cribb, M. W. Chase, and F. N. Rasmussen. *Genera Orchidacearum, Vol. 6: Epidendroideae (Part three)*. Oxford University Press, 2014.

USEFUL WEBSITES

World checklist of selected plant families (Orchidaceae)
apps.kew.org/wcsp/home.do
This has information about each published orchid species name and synonym and the geographical distribution for each accepted; there are also links to other online resources, such as Google images.

An online resource for monocot plants: e-Monocot
about.e-monocot.org
This has all the information from the *Genera Orchidacearum* series and can be queried by genus.

Internet orchid species photo encyclopaedia
www.orchidspecies.com
This has images of about half of all orchid species with habitat and cultural information.

World orchid iconography/Bibliorchidea
orchid.unibas.ch
This site contains the archives of the Swiss Orchid Foundation, including herbarium specimens, drawings, and images for over 11,000 orchids.

Epidendra: the global orchid taxonomic network
www.epidendra.org
This site has a variety of types of information and resources about orchids, including images, national park information, floras, and history.

First nature: nature and biology of orchids
www.first-nature.com/flowers/~nature-orchids.php
A website with a set of commonly asked questions about orchids with answers written by a non-biologist.

INDEX OF SPECIES BY COMMON NAME

INDEX OF SPECIES BY SCIENTIFIC NAME

ACKNOWLEDGMENTS

The illustrations of the orchids in this book are derived from botanical illustrations published in books and magazines during the nineteenth century.

Sources for these images, on which the illustrations in this book are based, are:

New York Botanical Garden
Pennsylvania Horticultural Society
Peter H Raven/Missouri Botanical Garden
Swiss Orchid Foundation at the Herbarium Jany Renz, Universityof Basel